THE TELESCOPE

THE TELESCOPE

ITS HISTORY, TECHNOLOGY, AND FUTURE

GEOFF ANDERSEN

PRINCETON UNIVERSITY PRESS
PRINCETON AND OXFORD

Published in 2007 by Princeton University Press, 41 William Street, Princeton, New Jersey 08540
In the United Kingdom: Princeton University Press, 3 Market Place, Woodstock, Oxfordshire OX20 1SY
First published 2006 in Australia and New Zealand under the title *Eye on the Sky* by Exisle Publishing Limited

Library of Congress Control Number 2006940308
ISBN-13: 978-0-691-12979-2
ISBN-10: 0-691-12979-7

British Library Cataloging-in-Publication Data is available
Printed on acid-free paper. ∞
press.princeton.edu

Printed in the United States of America

10 9 8 7 6 5 4 3 2 1

To my father, John

Acknowledgements

I have many people to thank for helping me with writing this book. These include Kathy, Jenny and John Andersen, Clark Chapman, Peter Foster, Martin Johnson, David Ottaway, David Shoemaker and Richard Cook for their comments. I would also like to thank Derek Buzasi, Richard Cook, Paul Davies, Dale Gay, Paul Hickson, Céline d'Orgeville, Chris Shelton, Eli Slawson, Gavin Rowell, Jack Wetterer, Ralph Wuerker and Eliot Young for their discussions and effort in getting the best images and photos for this publication.

Contents

Preface

In 2008 the telescope will be 400 years old. In that time telescopes have allowed us to experience places we will never be able to visit and to see sights we could have scarcely imagined. When most people think of telescopes they immediately tend to think of astronomers. These are people who have an enviable job in that they get paid simply to stare into space. Or at least that is the common perception. You may be surprised, then, to learn that the average astronomer has probably never even looked through their work instrument with their own eyes. Since the advent of photography and digital recording media, most astronomers can go through their entire observing careers without even needing to. Computer control even makes it possible to operate a telescope from the other side of the world – which in turn avoids working hours more suited to vampires. The dirty little secret is that, in many cases, the average astronomer has no better idea as to how their telescope works than the average Formula One driver understands the precise ins and outs of the car he drives. They don't need to – as long as they understand the operational constraints as they use it for their observations. Perhaps this book will help.

Most countries spend an extraordinary proportion of public funds on astronomy. There are many reasons, both scientific and social, for this. The social reason is perhaps the easier to understand. In general, the large investments prove to be rewarding in the field of public opinion – for one thing there are rarely any messy ethical or political issues for the funding agency. Press releases from observatories almost never contain bad news and most often the images presented are awe-inspiring. Since the dawn of time, humans have looked beyond our own tiny planet (even when they didn't picture it as such) with a certain amount of wonder. We have an innate desire to know what lies 'out there', how our existence began and what sort of a future (if any) we have, and these are all questions which can only be answered by looking past our atmosphere. Beyond this there is the manner in which we perceive our very existence; are we alone and are we unique? It is only by turning our attention outward that we can hope to answer these questions.

Intellectual curiosity notwithstanding, political realities ultimately decide what areas of research a country will choose to invest in. After all, it would be a stretch for even the best spin-doctors to justify the expense of a $200-million telescope solely on the basis that it may uncover deeper truths about a black hole located millions of light-years away. But then, to some extent the actual expense can be used as justification in itself. Amongst the sciences, astronomy is often (unfairly) seen as a Romance field of study – much like a geek version of supporting the arts. A country which invests significant amounts of money in this field is making a conspicuous display of wealth towards a purely intellectual pursuit and reinforces the perception of being civilised. This idea of a telescope as a status symbol is further compounded by the continual increases in observatory size. It has become a less than subtle game of one-upmanship amongst the richer nations which as much as shouts 'mine is bigger than yours'.

And increase in size they do. Nowadays a new facility will require the complete levelling of a mountaintop. Along with this comes an increase in operating costs – the leading facilities cost around $1 per second to operate. So why are they getting bigger? For that matter, why do we put them at the tops of inaccessible mountains in exotic locales such as Hawaii and Chile or in orbit around the Earth itself? Certainly, given the choice, who wouldn't want to go to work there? But we will see that there are actually quite practical reasons behind these locations. However, even if a telescope is to be located in the remotest part of the world, how can the costs get upwards of $1 billion for some proposed future instruments? What state-of-the-art engineering goes into them and just what do we hope to see?

The purpose of this book is to provide answers to these questions and to others. We will begin with a look at what can be seen with the naked eye. Beyond gaining an appreciation for this natural imaging system, we will examine its limitations. Next there is an historical account of the development of the telescope and the changes it created in our view of the universe. But more than being simply a litany of tales about ever-bigger telescopes, this book will describe, in simple language, how telescopes work and what new technologies are going into them. We will see how the same basic physics can be applied to state-of-the-art instruments on mountaintops and in space as well as to the store-bought system in your back garden. From there we will investigate many new techniques, such as adaptive optics in which computer-controlled deformable mirrors are being used to correct for the distorting effects of the atmosphere.

Most telescope texts are written by astronomers, so there is an understandable bias towards this application. However, investment in space-based surveillance easily matches that spent on civilian observatories. Spy satellites work on the same principles as all other telescopes, so it is a simple matter to figure out what they can see. We will also discover that although traditionally operated by secretive government entities, several commercial systems have recently become available to the private sector. Peering at the ground is one aspect, but space-based telescopes have also been constructed for defensive laser systems. Such 'Star Wars' lasers turned out to be impractical (or at least unaffordable), but a scaled-down version mounted in a jumbo jet is soon to be operational. Meanwhile, lasers of much lower power can be transmitted and received by telescopes for other purposes. Optical data transmission can provide us with high bandwidth communication to aircraft, satellites and interstellar probes. Laser ranging can be used to provide high-resolution topographic maps, sniff out chemicals or even monitor the distance to the Moon to within a few millimeters.

While its uses are quite diverse, the telescope is still a very simple device. The idea of this text is to present the many applications and engineering issues in a digestible manner. For the less technically minded, some of the basic mathematics and terminology is explained in more detail in the appendices. Included are rules-of-thumb to help separate the microns from the parsecs and the arcseconds from the electron-volts. Given the complex concepts explored here, it has often been necessary to generalise or simplify the explanations and numbers in order to make a readable but concise text. This means that in some cases an expert in the field may be tempted to shout, 'Well, that's not strictly the way it works!', and they'd probably be right. For example, I could say that a laser is a device which emits a single wavelength of light. Those familiar with the subject (including myself, I'll admit) may cringe and want to put in a dozen or so words of clarification, but in basic terms this statement is correct. In short, this book will help you understand what is going on but avoid, wherever possible, the gruesome details which keep so many of us employed.

One last note: a wide range of aspects of telescopes will be discussed in this book. In keeping with the emphasis on readability, however, the chapters are arranged by subject matter and not in any chronological order. If a particular topic is getting too technical, you can simply skip to the next one without running the risk of not being able to follow the later narrative. For example, the section on interferometry may be interesting to an avid amateur astronomer, but may not be digestible for everyone. Likewise, having finished the sections you are most interested in, you may return to the unread chapters at a later time.

Chapter 1

The naked-eye universe

Men at some time are masters of their fates:
The fault, dear Brutus, is not in our stars,
But in ourselves, that we are underlings.
– Cassius, from *Julius Caesar* (Act I, Scene ii)
by William Shakespeare

If you live in a big city, you have no doubt heard talk of stars. Next time you are driving across country at night or any time you are away from the city lights, stop for a while and take a good look at the sky. How many stars do you think you can see? Generally speaking, if you are in a dark enough location, you should be able to see around 1000 stars at any given time. As the Earth rotates throughout the night and stars rise in the east and others set in the west, you can increase your count to 2000, then 2001 when the Sun rises. The total number will grow further over the course of a year, with the slow motion of the Earth around the Sun. With a visit to other latitudes to see objects otherwise hidden below the horizon, the total would come to around 6000, depending on how good your eyesight is. Some of these 'stars' are in fact nebulae or even one of the five planets visible to the naked eye. The total number of objects is only slightly increased by throwing in one Moon and three galaxies. Including the odd comet and closely passing asteroid, the rare supernova, the occasional Earth-orbiting satellite, that's the entire list of objects beyond the Earth you can hope to see with the naked eye.

Given that our galaxy alone has some 10 billion stars in it, you might think that we are missing out on a lot. But our eye is doing a remarkable job even to present this view of our universe, and it is the result of much evolutionary modification. It has been theorised that the development of vision was responsible for the Cambrian evolutionary explosion some 540 million years ago. For around 3.5 billion years, life on Earth developed into just three phyla (a taxonomic grouping of plants or animals). Then, after a burst of evolutionary development lasting possibly as little as 5 million years, there were 38 phyla (which later extinctions reduced to the 35 we see today). It just so happens that natural light-sensitive receptors began to develop around this same period. Vision provides a huge selective pressure on animals to detect predators, prey, food, mates and the surrounding environment. Not surprisingly, then, eyes have developed in many different forms. Some animals are sensitive to ultraviolet or infrared radiation that we cannot see, while others have eyes specially designed for low light levels or polarised light. Even the different structures of eyes demonstrate a wide diversity: from the multi-lensed fly to the lobster, which produces images by reflection. Some creatures have zoom lenses, others have scanning optics while yet others have simple eyes which act like pinhole cameras. The eye is a remarkable piece of natural engineering, and in the case of humans, by far the most powerful of our senses.

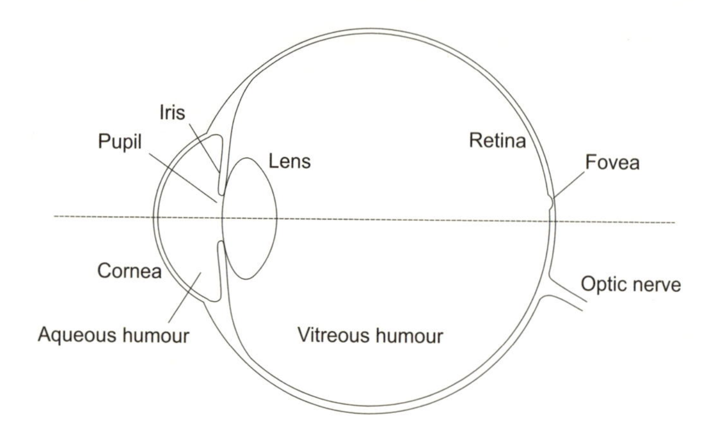

Figure 1.1: The human eye. Our best imaging takes place for light falling on the fovea which is populated mostly with colour-sensitive cones.

The human eye has many parts, but the basic design can be broken down into four major sections, as shown in Figure 1.1. There is a cornea for protection, an iris to alter the amount of incoming light and a focusable lens which forms an image onto the light-sensitive retina. The retina has two types of light-sensitive receptors, namely cones and rods. There are about 5 million colour-sensitive cones, predominantly located within a 1.5-mm diameter region called the fovea. These provide sharp imaging over a small field of view under good lighting conditions. The 100 million rods are distributed over the rest of the retina and while not able to distinguish colours, provide most of our peripheral and low-light vision.

Overall the eye is able to image at wavelengths of light (in effect defining the 'visible spectrum') from around 400 nm (violet) to 700 nm (red).[1] Individual cones are sensitive to a narrow range of colours centred approximately over the violet-blue, green and greenish-yellow portions of the spectrum. The actual colour of incoming light is inferred by the amount of activity triggered in each type of cone. For example, if the brain senses an equal output signal from neighbouring blue- and green-sensitive cones, then a colour lying midway between the two (something like aqua or turquoise) would be what the brain 'sees'. The cones are separated by about 2.5 micrometres (2.5 μm) in the fovea, so the smallest angle we can resolve under ideal conditions is around 1 arcminute (or around a thirtieth of the angle subtended by the full Moon). To put it another way, this means we should just be able to make out two distinct headlights on a car around 3 km away.

As a slight diversion at this point, it is often said that the Great Wall of China is the only man-made object visible from space. This is complete nonsense, as can easily be demonstrated. From the height of the International Space Station (350 km), an angle of 1 arcminute translates to 100 m on the ground. There are, of course, a multitude of structures larger than this. For example, the Great Pyramid of Giza, at 230 m on a side would be easily visible and appear quite separate to other pyramids nearby. On the other hand, the Great Wall is a little less obvious and most astronauts, including Chinese astronaut Yang Liwei, have said it was not visible. We will address this issue in greater detail later on, but the fact that the wall is quite long (over 6000 km) means nothing – the more important factor is that it is only 15 m wide. Under the right lighting conditions, it may show up against the background, just as we see stars at night but cannot resolve them.

This leads to another property of the eye which is truly remarkable – its sensitivity to a wide range of light levels. The eye's response to luminous flux is logarithmic, which permits us to accommodate both very bright and very dim scenes, often without even being aware of the difference. For example, you can comfortably read this book in direct sunlight, or by the light of a half-full Moon – a reduction in light levels by a factor of one million! By comparison, manufacturers of photographic film are happy with a film which can accommodate a range one thousand times smaller. Even the most advanced digital sensors would find it difficult to accommodate these extremes. Of course, the human eye cheats a little by using cones for bright light and rods for low light levels. For astronomy then, rods are the more important detector in the retina, and they lie outside our direct field of vision (the fovea). This means that to see the faintest stars it is often best to use averted vision. That is, you should look to one side of the star so that the light does not fall on the fovea, which is predominately populated with less sensitive cones. The problem with rods, of course, is that they have no colour sensitivity. This is fairly evident when you look at a nighttime scene dimly lit by the Moon, where everything loses its colour and the scene appears in shades of grey. The colours are still there – you simply can't see them.

Beyond just collecting light, the human eye forms images of distant objects. These images are produced on the retina by the lens, in precisely the same way as a camera lens forms images on film. We'll come back to the concept of imaging later, but for now we should note that by changing the tension in muscles surrounding the lens of our eye, the shape of the lens can be stretched or compressed in order to maintain focus over a wide range of viewing distances. The retinal image is actually produced upside down, but the brain does its own correction to this unreal situation and inverts everything back to normal. The brain then combines the light from two eyes which look at a scene from slightly different angles (parallax) to give us a sense of three dimensions. In summary, the human eye can be used to detect colours, shades, shapes, dimensions and distances. From an engineering point of view, the eye is a truly remarkable instrument; even more so given that it is the result of millions of years of random trial and error.

Of course, the human eye does have some limitations. We cannot see radio waves, microwaves, infrared, ultraviolet, X-rays or gamma-rays. The refresh rate of the visual signal processing (called persistence) is around 20 times a second and changes occurring faster than this cannot be seen. At the same time, we cannot brighten the image of a dim object by staring at it for a long period of time in the same manner as a time-lapse exposure on film. Another difference between the eye and cameras is that there is no way of recording our retinal views on a permanent medium for others to see. The eye is susceptible to fatigue, disease and aging, which can affect resolution, focusing and sensitivity. On top of this, the eye (or more correctly, the brain) can also be confused and fooled by certain arrangements of objects. In spite of these limitations, though, the eye is a remarkably versatile instrument.

Often when considering a telescope, people will be prompted to ask: 'Just how far can it see?' Later in this book it will become clear just how meaningless this question is, but for now consider applying the same question to our eye. You can see objects sitting right in front of your face, but at the same time you can see a mountain dozens of kilometres away. Alpha Centauri is around 40 trillion kilometres away, and Andromeda Galaxy, also visible to the naked eye in dark skies, is nearly a million times more distant than this. So really, whether the eye can see an object comes down to how big and bright the object is, not how far away it is. Of course, seeing light from a star is not the same as forming an image of it – after all, a point of light is not an illuminated disk, so this is where it becomes meaningless to consider these sorts of issues. In fact, it is because we cannot see any of their details that cosmic objects have always held such a fascination for us.

As you look at the stars, try to imagine yourself as an Egyptian sailor on the Mediterranean Sea some four-and-a-half thousand years ago. From an early age, you would have been taught how to find the dim star Thuban. Throughout the entire night, it would serve as an unwavering beacon by which to navigate. Even after the most disorientating of storms, this star always lay to the north. It is no wonder, then, that Pharaoh Khufu aligned his Great Pyramid to this most important of stars. It is in contemplating the history of naked-eye observations of the sky that we begin to appreciate why such observations have had such a powerful effect on mankind throughout the ages.

Every star (except the Sun) rises and sets around four minutes earlier every night. Over 365 days, this amounts to a complete day, so the rising or setting of a particular star at a particular time can be used as a measure of a year. The same Egyptian sailor, as a long-time observer of the heavens, would know that the rising of certain groupings of stars at dusk indicates the onset of particular seasons. This knowledge could be used to anticipate the annual flooding of the Nile and for planning harvests and plantings. To nomadic civilisations, accurate timekeeping was equally important for following migratory animals and preparing for the harsh environmental conditions of particular seasons. As civilisations developed, crop rotations and agricultural storage requirements were determined by a 'calendar' which was essentially a measure of the motion of the stellar sphere.

Naturally it was easy to believe that these groupings of stars were not simply passive markers of time, but exerted an active influence over life on Earth. These groupings (asterisms or constellations) were associated with physical objects or gods as long ago as 4000 BCE, and from there it was a small step to anthropomorphising their forms. The constellations thus became powerful gods influencing our lives, which inevitably led to religious and superstitious rituals and beliefs. Temples were built and offerings and sacrifices were made to influence the gods to provide bountiful harvests and improve living conditions. These gods were regarded as powerful enough to control human existence, so they were not to be trifled with. The intertwining of superstitions and religion continued even into the modern era. For example, the Catholic Church used astrological charts until well into the eighteenth century.

While many people still believe in astrology today, it clearly makes no sense. The constellations used for astrological 'signs' are merely apparent groupings of stars which are actually unrelated and lie at completely different distances. They have no physical existence – you could never 'visit' a constellation. Furthermore, there are three planets which are used in present-day astrology whose existence was not even known before the eighteenth century, so those who purport to be using 'ancient mystical knowledge' are relying more on the gullibility of the uneducated than on any secrets of civilisations past. It is also worth mentioning that there are in fact 13 Zodiacal constellations, with Ophiuchus lying between Libra and Sagittarius.

While the stars appear to move around us, their motions are really only telling us where we are and what time it is.

The Chinese, well aware of the 'clockwork' nature of the celestial sphere, used the apparent motions of the Moon and the Sun as the basis of their calendar. A Chinese year can have 353, 354, 355, 383, 384 or 385 days on a cycle which repeats every 60 years. It may seem a little confusing and cumbersome at first, but it works. Proof of this is the fact that the Western year 2006 CE is year 4703 in the Chinese calendar. In fact, it is the longevity of this calendar which has made it possible for historians to examine Chinese records to determine accurately the precise dates of ancient events, both celestial and social. Likewise, the Jewish and Muslim faiths have, for centuries, used the cycles of the Moon for daily observances and calendrical measurement. The Hebrew calendar date for 2006 CE is 5766 AM (*Anno Mundi* – in the year of the world). By comparison, the Gregorian calendar seems positively newfangled. It also puts into perspective all the superstitious fuss that went along with the dawning of the third millennium.

I was once observing a transit of Mercury across the Sun with a student who remarked on the amazing punctuality of the event. Quite the opposite, I pointed out, she should be more amazed that our clocks are timed with such precision to celestial events. After all, this is where we get our system of time measurement. In fact, measurements of the passage of certain stars across a 'fixed' point or line in space were used to determine the length of what would become our Gregorian calendar year. While the actual tropical year is determined by the Sun, a rough approximation can be made by measuring the time between successive transits of a given star across an arbitrary north-south line in the sky at a certain time of the day. For example, if you have a telescope aligned north-south, and Vega is centred in your view at exactly 9 pm, then it will be centred again at the same time a year later.

An example of the clockwork precision of celestial motions was never more evident than at the turn of the sixteenth century. While on his fourth voyage to the New World, Christopher Columbus became stranded in Jamaica. At first, Columbus and his crew were well received by the locals, who fed and housed them while they waited for a rescue party. However, as the months passed, he and his badly behaving crew became increasingly unpopular with the indigenous population. Relations degraded to such an extent that there was a distinct possibility of a bloody confrontation. Sensing this, Columbus invited the chiefs to a meeting with him on 29 February 1504. Columbus stated that God was unhappy with their treatment, and threatened to remove the Moon from the sky. That night, as Columbus well knew, there was to be a lunar eclipse. Sure enough, as foretold, the Moon began to darken and redden. The natives, understandably impressed and more than a little terrified, begged Columbus to get his God to return the Moon. Columbus 'pondered' their request for a while and eventually agreed to do so (he could afford to play it slow as a lunar eclipse can last for hours). From this point onwards the locals pretty much did his bidding until the rescue party arrived several weeks later.

Astrometry, the measurement of stellar positions, is also essential for navigation. Throughout the course of the night, the stars in the sky will rotate about the celestial pole. In the Northern Hemisphere, this point in the sky is very close to Polaris (the North Star or Alpha Ursae Minoris). Taking a long-exposure photo of the night sky towards either pole will demonstrate this, as shown in Figure 1.2. The Earth also has a slight wobble, which causes its axis of rotation to precess (describe a circle) over a period of 25,800 years. Around 4500 years ago, the north celestial pole lay near a star called Thuban, but today it is close to Polaris. In fact, precession will continue to narrow the angle between them until 2106. But even now, Polaris serves as a much better guide to true north than magnetic north measured by a compass. This relationship is quite fortuitous (and unlikely) and in the Southern Hemisphere there is no bright star close to the south celestial pole.

Figure 1.2: Star trails around the south celestial pole. Over the course of evening (and the day too), the stars appear to move around the celestial poles. A time-lapse photo shows this motion as the stars form circular star trails. The shorter trails are nearer to the pole. Courtesy: Anglo-Australian Observatory.

While navigators can get their bearings (compass orientation) from the celestial poles, knowing your current location on the Earth is also critical for navigation. For this we can use the celestial poles once again. Quite simply: the angle of the pole above the horizon is your latitude. Live in Oslo? Since your latitude is 60 degrees North, Polaris is 60 degrees above the horizon, or a third of a way around the sky. Longitude is much trickier to determine. In order to know how far you are east or west of the Greenwich meridian, you essentially need two clocks – one telling you the local time, and one telling you the time it is in London. For each hour's difference between the two clocks, you are a further 15 degrees removed from zero degrees longitude. So, if your local time is four in the afternoon and it is noon in London, then your longitude is 60 degrees East.

For millennia, navigators had been able to measure latitude, but it was not until towards the end of the eighteenth century that John Harrison developed a sufficiently accurate timepiece for longitude measurements at sea. For accurate navigation today, we mostly use the Global Positioning System that consists of 24 or so satellites in precisely known locations. GPS uses the relative timing of signals from atomic clocks on these satellites to determine position (latitude, longitude and even altitude) on the ground. These days there are international organisations that exist to take care of the details and help us better define time and location to the nth decimal place. Before the invention of the telescope, however, people had to rely on less scientific authorities to determine such things in their lives.

At the turn of the sixteenth century in Europe, the Catholic Church had pretty much determined the way things were in the universe. Following a strict Ptolemaic doctrine, there were several assertions which were not up for debate. The Earth was the centre of the universe (by virtue of it being the ultimate work of God), and the planets and Sun all rotated around it in circular orbits. All bodies in the universe were perfectly spherical and except for the odd comet, the celestial sphere was permanent and unchanging. Over much of the Dark Ages, this simple arrangement was immutable. But as measurements of planetary positions improved, adjustments had to be made to make doctrine agree with reality. For example, the planets simply refused to move in precisely circular orbits about the Earth. In order to correct for this, some planets were permitted to travel in small circles about a point which itself moved in orbit about the Earth. This stop-gap measure worked up to a point, but it was becoming clear that further improvements were required. So it was that in 1540 Nicholas Copernicus published *De Revolutionibus* with the primary tenet being that the Sun, not the Earth, was the centre of the universe.

The Sun-centred model solved many nagging problems associated with planetary motions but was still far from perfect and often required as much tweaking as had the Earth-centred model. Fortunately for Copernicus (depending on how you look at it) he died within days of the publication of this work, and so did not have to answer for any of the subsequent controversy. While it was not widely read, the main thesis was discussed widely around Europe. For the most part, the Church largely ignored it or at least treated it as an annoying work of fiction. After all, there were no gross errors in the motions of nighttime objects that could seriously threaten the basic canonical law and besides, the Sun-centred model was not new, but had in fact been suggested by Aristarchus (and reported by Archimedes) at some time during the third century BCE. Since the Church was able to ignore this for nearly a millennium, the resurgence of the idea was no great cause for concern.

Tycho Brahe[2] is perhaps one of the last of the pre-Enlightenment 'scientists'. His life work consisted almost entirely of observations and cataloguing of phenomena without much effort to devising an underlying theoretical framework to describe the 'why' behind the phenomena or making predictions for testing these theories. However, this should not diminish his achievements in any way – far from it. In fact, his observations were of such a high quality that they were used by his assistant to devise the first laws of motion of planetary bodies, which are still used today. And he did all this without the aid of a telescope.

Brahe was born in Skane, Denmark (now in Sweden) and raised by his aristocrat uncle in an environment which emphasised education and critical study. Early on, perhaps as the result of observing a partial eclipse in Copenhagen, Brahe became intrigued with celestial observations and the ability of astronomers to predict such events well in advance. Throughout his early life, he studied at various universities, eagerly pursuing a future in the sciences. In his day, most astronomy was actually astrology, and Brahe made a name for himself when he used a lunar eclipse of 1566 to correctly 'predict' the death of Suleiman the Magnificent, ruler of the Ottoman Empire. Since Suleiman was 72 years old, it may be said that this 'prediction' was not much of a stretch, and it was even less impressive when it was later discovered that Suleiman had actually died a few weeks before the eclipse – but word of this had failed to reach the Danish court. A further significant event in Brahe's early life occurred when the tempestuous young man got into a duel that resulted in him losing a portion of his nose. For the rest of his life he covered up this disfigurement with a prosthesis of gold and silver. Much of his early work is fairly unremarkable, except perhaps for the fact that his studies gave him a good knowledge of the heavens. This would prove to be instrumental in a discovery that would bring him his fame.

On the evening of 11 November 1572, Brahe noticed a new star in the Constellation Cassiopeia. Over the next few months, he continued to observe the star and eventually published his findings in *De Nova Stella* (About The New Star). While others had observed the star, Brahe made the bold assertion, on the basis of his measurements, that this was indeed a new object and not a new comet or meteor which were the only other known objects that could appear in the sky. His revelation should be viewed in the context of the time in which he lived. It has already been mentioned that the Catholic Church had held steadfast to the view that the heavens were perfect and unchanging. Brahe's new star challenged this prevailing doctrine and could not be ignored. Such was the impact of this work that today we still refer to similar events as 'novas' (and supernovas, which is in fact what Brahe's new star was). For Brahe personally, the acclaim the work received had more immediate financial benefits, when he managed to parlay it into a Royal tenure. The King of Denmark gave him a generous income and support for the creation of Uraniborg observatory, which Brahe built on Hven, an island near Copenhagen. While not quite what we think of as an observatory in today's terms (as there was no telescope), it was a site for observing and documenting the heavens in a consistent manner.

To aid his observations, Brahe constructed a fine quadrant which could measure the altitude of stars above the horizon to great precision. For the rest of his life he made copious measurements of the positions of celestial objects and it is for these measurements that Brahe should be noted. While to most people they resemble a tedious catalogue of numbers, they far surpassed the accuracy of any previous measurements. Most were good to a couple of arc minutes, and some to better than a quarter of this – something which can be compared to the size of the printed words on this page on the other side of a large room. It was as a direct result of these unprecedented observations that his assistant, Johannes Kepler, was able to formulate his laws of planetary motion.

Kepler was a long-time supporter of the Copernican Sun-centred universe theory but lacked the eyesight for the high-quality observations required to evaluate the theory. Over many years he sought to work with Brahe (or rather, his data) and was eventually hired as his assistant early in 1601. Brahe was very protective of his measurements and would only allow Kepler to look at limited portions of it. This all changed when he died later that year and Kepler inherited the valuable catalogues. With his gift for mathematics, Kepler was able to use the data to formulate three laws of planetary motion (published in 1609) which went a huge way to legitimising the Copernican universe. The three laws are:

1. The orbits of planets are ellipses, with the Sun at one focus of the ellipse.
2. As the planet moves in orbit about the Sun, the line joining the planet to the Sun sweeps out an equal area in an equal time.
3. The ratio of the squares of the orbital periods of two planets is equal to the ratio of the cubes of their semi-major axes.

The underlying physical reason for these laws being the way they are was not really understood until Newton formulated his Theory of Gravitation some 50 years later. However, these simple laws made it possible to accurately predict the motion of planets, and are still used today for all but the highest precision orbital calculations. Kepler's work was incredibly important, and began a gradual change in opinion away from the Earth-centred universe theory, and from the unquestioned acceptance of religious dogma. Alone it would have caused an upheaval in the way the world viewed the universe. However, even as it was being prepared for publication, an optical worker in Holland had invented a device which would dramatically increase the pace of inquiry into the physical universe and usher in the Enlightenment.

Chapter 2

The development of the telescope

As a result of the lack of documentation and because records have been lost in the course of time, the invention of the telescope cannot be simply attributed to a single person. The first description of any sort which could concern a telescope comes from Thomas Digges of Oxford who, in 1571, describes a device built by his father (Leonard), who:

> ...was able, and sundrie times hath, by proportionall glasses duely situate in convenient angles, not onely discovered things farre off, read letters, numbred peeces of money with the very coyne and superscription thereof, cast by some of his friends of purpose upon Downes in the open fieldes, but also seven myles off declared what hath been doon at that instante in private places.

It would seem that this describes a device requiring more than one lens and providing a magnified view of distant objects.[3] If this was indeed a written description of the first telescope, it remains a mystery as to why the device did not gain more attention or further development. In fact, many historical snippets have been found which could hint at other inventors of the first telescope, including Roger Bacon. But again, the question of why they didn't become more widely known remains. In fact, the credit for its invention would go to a Dutchman some 37 years later.

Hans Lippershey was a spectacle maker living in Middelburg, Holland when, as one story goes, some children entered his shop and began playing with various lenses. One of them noticed that holding a certain combination of lenses at the right distance in front of the eye gave a magnified image of distant objects. Lippershey supposedly did some tinkering and found a useful combination giving an acceptable magnification, but given the long focal lengths involved in the final instrument, this story is dubious at best. However, what we do know is that on 2 October 1608, Lippershey was awarded a patent for a device which is the same as a modern-day refractor telescope. From other records it is clear that he built several devices for the military and laboured hard to keep the precise design a carefully guarded secret. This ended up being to his detriment in the annals of history as he did not gain nearly the recognition he probably deserved. Most people erroneously attribute the invention of the telescope to Galileo Galilei.

Galileo Galilei was born in Pisa, Italy on 15 February 1564 – the eldest of six children of Vincenzo Galilei, a musician. Home-schooled at an early age, he was encouraged to study medicine as a profession offering good financial security. Throughout his studies he displayed a highly free-thinking and sceptical nature which often raised its head in lectures, where he readily questioned accepted knowledge. During his time at the University of Pisa he realised an attraction to mathematical sciences and began attending lectures on the subject. In the end he neglected his medical studies to such a degree that he finished his years at the university without receiving any formal degree.

It was during this time, however, that he made his first observations of the motion of pendulums. Supposedly, while in church he noticed that the time for a swing of a chandelier remained constant even as the size of the arc decreased. In research he conducted later he showed that the swing period was dependent only on the length of the pendulum, not on the mass of the bob or the initial release point.[4]

Figure 2.1: Galileo Galilei, on the 2000 lire banknote circa 1973.

Through the intervention of an influential contact, Galileo took up a mathematics professorship at the University of Pisa. During this time Galileo demonstrated the fallacy of the Aristotelian view that objects of different mass fall at different rates, though it is unlikely he did so using balls dropped from the Leaning Tower of Pisa. In 1592 he moved to a better paying job in Padua (Padova), where he would spend the next 18 years of his life. There is much to tell of the life of Galileo – too much to go into detail here. At this point I will merely say that, as a result of his work, he became well known to the Venetian court and well respected amongst his peers, even while he demonstrated a total lack of tact and tolerance towards those who disagreed with his views. In fact, his outspoken nature would end up becoming an important factor in how his life played out.

We return, then, to Galileo living in Padua in mid-1609, when word reaches him concerning a new invention which could magnify distant objects. While tapping his friends for more information and trying to procure an actual device to study, Galileo was informed that a Dutchman had arrived in Venice (Venezia) with the intent to sell a spyglass to the Doge (the city-state ruler). Galileo did not want to miss a critical financial opportunity such as this, so he decided he had to build one of his own. In the meantime, a good friend of the Doge delayed the evaluation of the Dutch spyglass to buy Galileo some time. Armed only with vague, second-hand reports of how the instrument operated and how it was used, Galileo began the process of divining its design and construction. In perhaps the single most impressive feat of reverse engineering in history, he took just one day to work out the design and improve on it to produce a device far superior to anything in existence at the time.

We are not entirely sure of the design of the Lippershey telescope – there are two possible lens configurations, as we will see in the next chapter. However, we do know that Galileo's system consisted of a concave and convex lens – a combination we today call the 'Galilean telescope'. Over the next few weeks he conducted experiments to improve the design. The telescopes were very long (up to a metre or more) and had very small lenses. With adaptations to the lenses, Galileo managed to increase the magnifying power well beyond that of competing designs – up to 20 times compared with the prevailing value of around five.[5] When Galileo presented his telescope to the Doge (as a gift) he was immediately awarded tenure and a huge pay rise. Even if that were the end of the story, Galileo would warrant a minor section in any history of the telescope, but this was merely the beginning. With his financial security assured, he was quick to use the telescope to study the world around him.

Galileo first turned his spyglass on the Moon, where he observed that the dividing line between the light and dark sides (the terminator) was ragged and rough. He was quick to draw the conclusion that the surface was pock-marked with ridges and depressions and even accurately described a prominent crater near the centre (believed to be what we now call Albategnius crater). He then went further and made calculations of the heights of the mountains and estimated them to be up to five miles high (which is now known to be about two to three times too large an estimate). This seemingly simple conclusion was at odds with the prevailing wisdom at the time that all celestial objects were perfect spheres.

Galileo then turned his attention towards stars that, he noted, still appeared as points of light with no discernible detail. He described the appearance of many more stars invisible to the human eye and discovered that the Milky Way gets its appearance from a multitude of stars. This discovery may seem unimportant to us, but it has to be examined in terms of the thinking of the day. While today we have no trouble believing in the existence of things we can't view with our naked eye, the telescope was the first device that presented objects previously invisible to unaided senses. Even more sensationally, these new objects had no precedent in religious scripture, so it was difficult to say whether they were permitted doctrine or not. As if that were not enough, the real impact was to come when Galileo made observations of Jupiter.

Galileo first viewed Jupiter on the night of 7 January 1610, and he was surprised to notice three bright points of light next to the planet, all forming a straight line. It was the next night that he got his real shock, when he discovered that the positions of these points of light had changed with respect to the planet, while still remaining in a straight line. Over the next few months he made many different nighttime observations and documented the new positions each night. Galileo made the leap of intuition that the points of light were in fact objects (which he called the Medician planets) in orbit around Jupiter.[6] An example of his descriptions, taken from *Sidereus Nuncius* ('The Sidereal Messenger') can be compared with calculations made with modern-day software as shown in Figure 2.2.

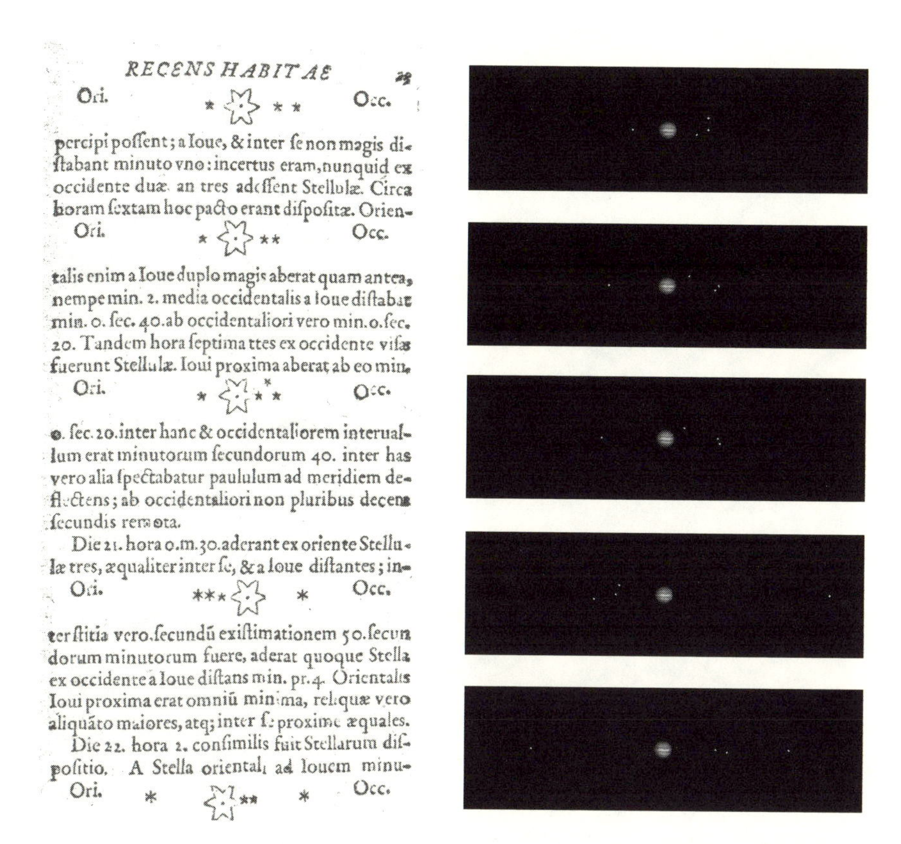

RECENS HABITAE 29

Ori. * * * Occ.

percipi poſſent; a Ioue, & inter ſe non magis diſtabant minuto vno: incertus eram, nunquid ex occidente duæ an tres adeſſent Stellulæ. Circa horam ſextam hoc pacto erant diſpoſitæ. Orien-

Ori. * ** Occ.

talis enim a Ioue duplo magis aberat quam antea, nempe min. 2. media occidentalis a Ioue diſtabat min. 0. ſec. 40. ab occidentaliori vero min. 0. ſec. 20. Tandem hora ſeptima tres ex occidente viſæ fuerunt Stellulæ. Ioui proxima aberat ab eo min.

Ori. * * * * Occ.

0. ſec. 20. inter hanc & occidentaliorem interuallum erat minutorum ſecundorum 40. inter has vero alia ſpectabatur paululum ad meridiem deflectens; ab occidentaliori non pluribus decem ſecundis remota.

Die 21. hora 0. m. 30. aderant ex oriente Stellulæ tres, æqualiter inter ſe, & a Ioue diſtantes; in-

Ori. *** * Occ.

terſtitia vero ſecundũ exiſtimationem 50. ſecundorum minutorum fuere, aderat quoque Stella ex occidente a Ioue diſtans min. pr. 4. Orientalis Ioui proxima erat omniũ minima, reliquæ vero aliquãto maiores, atq; inter ſe proxime æquales.

Die 22. hora 2. conſimilis fuit Stellarum diſpoſitio. A Stella orientali ad Iouem minu-

Ori. * ** * Occ.

Figure 2.2: A comparison of sequential observations of Jupiter and calculations of the positions of the moons using astronomical software (with the moons enlarged relative to Jupiter for clarity).

Sidereus Nuncius caused an immediate sensation, not just in Italy but around the whole of Europe. Galileo had presented observations of phenomena which seemed to disagree with prevailing Church doctrine that the entire universe revolved around the Earth. Furthermore, the lay person need no longer simply accept the Church's view of the universe – anyone could pick up a telescope and see for themselves. Galileo published several more works describing observations of other celestial objects, such as Saturn. Originally he believed the latter's odd appearance was due to two close satellites on either side, but in later drawings he depicted it in many ways, including surrounded by rings as it is now known to be. However, he never conclusively stated any preferred view on the matter. That discovery would be attributed to Christiaan Huygens, whose work we will discuss later on.

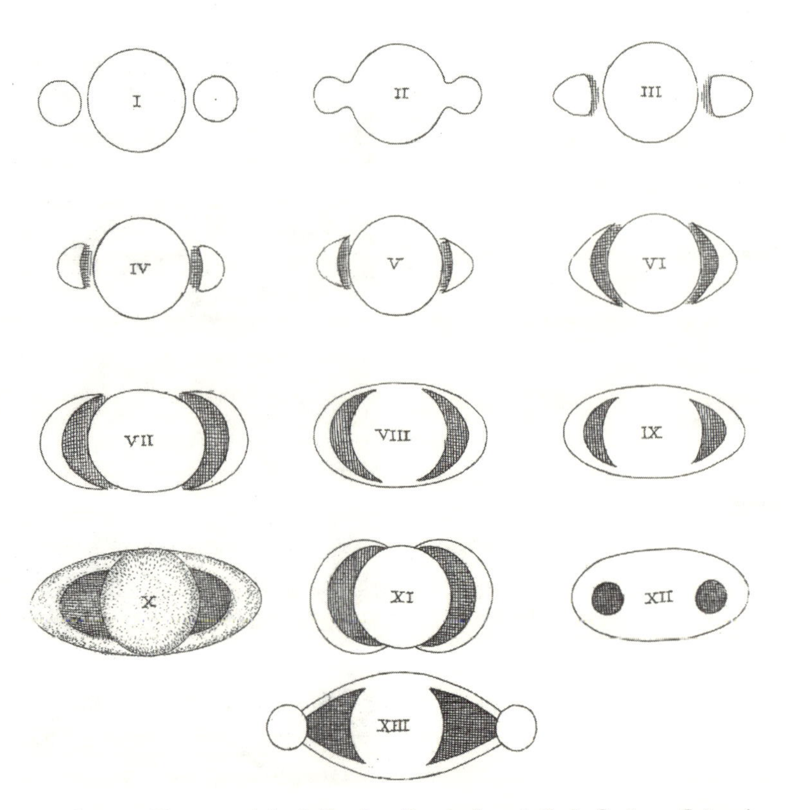

Figure 2.3: Images of Saturn made by Galileo from *Brevis Annotatio in Systema Saturnium*, 1660. It is interesting to note here how the lack of resolution made it difficult for Galileo to get an accurate idea of the nature of these rings.

Galileo made numerous observations of sunspots, though he wrongly tried to take credit for discovering them. He was, however, the first person to theorise that the blemishes were in fact on the surface of the Sun (and were not objects passing in front of the disk), thus removing one more object from the list of supposedly perfect celestial spheres. A more significant series of observations came when Galileo documented the appearance of Venus. Galileo observed that not only did Venus change in apparent size, but it also changed in shape, moving through phases much like the Moon. The identification of phases served as the final nail in the coffin of the Earth-centred view of the universe. Venus cannot move through a complete set of phases according to the geometry of Ptolemaic theory, but can under the Copernican model. Publication of the inflammatory *Dialogue Concerning the Two Chief World Systems*, in which Galileo espoused his views on the Sun-centred universe, eventually put him in direct conflict with the Catholic Church. Inevitably he was brought before the Inquisition and forced to publicly recant and declare that he was wrong in teaching the heretical theory as fact. As part of the plea bargain, his execution was commuted to a sentence of house arrest, under which he spent the remaining eight years until his death in 1642.

The semantics of this event are perhaps telling. Rather than demanding that Galileo admit that the Sun-centred theory was wrong, the Catholic Church deemed his mistake to be in teaching the theory as a true description of reality. The Copernican/Keplerian model of the Solar System was already proving far superior to the Ptolemaic theory in its power to predict the motions of the planets and was being widely used in all fields from navigation to astrology. No doubt the Vatican could already see this success as an indication of the fact that it would remain in use, but still held on to the idea that its interpretation of scripture was the one true reality. This led to the Vatican's chosen method of prosecuting Galileo. The Church further placed *Dialogue* on its official list of banned books – where it remained until 1882. In 1992 Pope John Paul II rescinded the edict of Inquisition against Galileo – a mere 350 years after his death. Today, if you visit the Istituto e Museo di Storia della Scienza in Florence you can see some of his telescopes and, disturbingly, one of his fingers.

Galileo would have no trouble in recognising his refractor telescope in a shop today. Although he was not the inventor of the telescope, the controversial aspects of his interpretations of what he saw with his instrument are widely known. Equally sensational was the great stir the telescope itself caused throughout Europe. One reason for this is that it was a passive device that completely altered the way warfare was conducted – after all, if you can see ships or armies at a much greater distance, you have more time to prepare for attack or defence. The other reason was a deeper one; here was a device that made things visible that were otherwise invisible to the naked eye.[7] The telescope (and later the microscope), were thus two major devices that helped usher in the Enlightenment. Suddenly, anyone could experience things beyond the range of the unaided human senses, and start questioning conventional wisdom about the universe in which we live.

While Galileo was able to make and improve upon the design of the telescope, his understanding of its operation was mostly in engineering rather than physical terms. That is to say, he did not know *why* certain lens combinations produced the resulting magnification, only *that* they did. In fact, at this point in history, no one yet understood how a single lens produced an image, let alone how combinations of them operated. That was until a Frenchman, René Descartes, took up the problem. A Jesuit-educated lawyer born in 1596, Descartes, like Galileo before him, abandoned his chosen profession in order to pursue studies in mathematics and sciences. He is often remembered for establishing the Cartesian coordinate system – a way of representing the position of any object in space with respect to three mutually perpendicular axes.

Descartes also advanced many ideas which laid the foundation for much of the later work done by Newton. Most importantly for this story though, in 1637 he showed how a lens focuses light and how images are formed. He also described how some imperfections in the construction of a lens could degrade image quality. This in turn led to methods for constructing better telescopes. Descartes made a further, indirect, contribution to telescopes by inspiring a young Dutchman, Christiaan Huygens,[8] into a career in physics. Huygens began with investigations into chromatic aberration – a problem with lenses which causes white light images to become blurred. Along with his brother, he was able to capitalise on this theory to create vastly improved telescopes. He then used one such telescope in 1655 to discover the first moon in orbit around Saturn (Titan). A year later, Huygens elucidated the true nature of Saturn's elongated shape as being due to flat rings encircling the planet.

In 1678 Huygens developed a wave theory for light, which he did not publish until 1680. This work was a culmination of years of work in optics, and also relied on a measurement of the speed of light made by a Dane, Ole Rømer, in 1671. Along with Cassini (famous for identifying a gap in Saturn's rings which bears his name), Rømer developed a particularly elegant method for making the measurement. Jupiter periodically eclipses its moons according to a predictable schedule, so Rømer reasoned that if the velocity of light was finite, then there should be a delay in the disappearance of a moon behind Jupiter's disk at times when the Earth was further away from Jupiter due to their respective orbits around the Sun.

With high-precision timings of these delays, Rømer was able to calculate a velocity of 232,000 km/s. This may not seem too close to the value of 300,000 km/s that is used today, but Rømer's calculation was in error largely because of insufficient knowledge of the size of the planetary orbits. The measurement is important as it represented the first time that celestial objects, rather than equipment on Earth, were used to pin down a physical quantity. It is extraordinary to contemplate just what sort of changes had taken place in little over half a century. The telescope was initially viewed with suspicion, with many people questioning the validity of 'seeing' something which could not be observed with the naked eye. The acceptance of Rømer's measurement shows that by this stage most people had relinquished the requirement of physical interaction with something in order to accept it as real.

At this point in history we come to Sir Isaac Newton. Countless books have been written about this pre-eminent physicist and, once again, there is insufficient space in this book to recount all of his achievements and discoveries. It could be said that his Theory of Gravitation and co-discovery of Calculus were important to astronomy, but they are of little relevance to this text. Sticking to the story at hand, we must limit ourselves to his contributions to telescopes. Even so, there is still an impressive amount of material to cover.

Newton was born on Christmas Day, 1642 – the same year that Galileo died. Something of a loner (perhaps due to a fairly unpleasant disposition), Newton did much of his seminal work in optics during a period of reclusion in the English countryside while avoiding an outbreak of the plague in London. In 1704 he published a summary of his findings in *Opticks*. Amongst the important findings detailed in this text are the following:

1. White light is made up of the complete spectrum of colours.
2. The colours can be separated out by passing the light through a prism, as different colours are bent (refracted) at slightly different angles.
3. This 'dispersion' (not his word for it) is the cause of the chromatic aberration that Huygens found to be affecting refractor (lens-based) telescopes.
4. There is no dispersion when light is reflected.
5. Telescopes could be improved by using a reflecting focusing element (a mirror) instead of a refracting one (a lens), since all colours will focus at the same point. This type of telescope is now known as a Newtonian reflector.

It was clear from this work that reflectors would give superior performance over refractors. Still, lens-based telescopes remained the common instrument until mirror technology improved to a point where the benefits could be exploited. For example, mirrors in Newton's day were mostly produced by polishing a metal surface, often tin, silver or speculum[9] (an alloy consisting of four parts copper to one part tin). These mirrors all suffer from the problem of oxidation (tarnishing), which means that the surface has to be repolished regularly. Even when superior mirror-making methods brought about the dominance of reflector telescopes, refractors continued to be used in non-astronomical applications where a slight amount of chromatic aberration could be tolerated. Even today, the cheapest telescopes sold in stores are refractors. The last great astronomical refractor was the Yerkes telescope, with a diameter of 1 m, built in 1897 for the University of Chicago.

Like all scientists, Newton made his fair share of mistakes, which we are able to identify with the benefit of hindsight. Unfortunately, such was his prestige that his opinions carried much weight and seriously delayed progress in some areas. Most notably, Newton firmly believed that light was corpuscular, i.e. it came in particle form. In spite of the obvious predictive power of Huygens' wave theory of light, introduced some 50 years earlier, the force of Newton's opinions cast a long shadow. Even well after Newton's death in 1727, the general consensus was to lean towards the distinguished Englishman's view simply because of his reputation alone. However, there was one phenomenon actually discovered by Newton which was to swing the debate, even though this would not happen until the end of the eighteenth century.

In *Opticks*, Newton described an experiment where a slightly convex glass plate was laid on top of a flat glass plate. When lit with white light, very small colourful rings could be seen encircling the point where the plates touched. Under monochromatic light (light of a single colour) the fringes appeared alternating light and dark. The colours seen when there is a thin layer of oil on a wet road are the result of the same phenomenon. These rings are today called Newton's rings, and the discoverer himself had a hard time trying to explain their nature. In the first years of the nineteenth century, a man called Thomas Young developed another arrangement where this 'interference' effect could be seen. Careful analysis by Young led him to explain the fringes as being due to the combination of light waves. We will return to the matter of interference (and Thomas Young) in a later chapter. The important point here is that the wave theory of light had been developed by Huygens to explain refraction, and Young was able to use it to interpret interference. Public opinion about the nature of light was beginning to shift a little. The final proof came with Fresnel and his theory of diffraction.

Diffraction is the bending of light around an obstacle or through an aperture. Some diffractive effects can be seen in our everyday lives, but are often difficult to observe. If you squint heavily at a light bulb, you may notice that a flare of light appears to stretch out from the top and bottom of the globe. This spreading of light as it passes through the narrow slit of your eyelids is diffraction.

In 1817 the French Academy of Sciences offered a prize for the best explanation for the phenomenon of diffraction. One entry came from a Frenchman, Augustin Fresnel, in the form of a long treatise expounding the wave theory of light. The theory was not too popular with Laplace, Poisson and Biot, members of the judging panel who favored the corpuscular theory of light. Failing to find any mistakes in the mathematics, Poisson laboured hard to finally find what he thought was a conceptual error. Quite simply, if light spread out when passing through an aperture, then it could equally spread around the edges of a barrier to appear on the other side. This meant that if you placed a small round disk in the path of a beam of light, you should expect to see a bright spot downstream in the middle of the shadow. Clearly this was preposterous, so the theory must be wrong. Fortunately for Fresnel, he had an advocate in the chairman of the committee (François Arago), who insisted that the experiment be carried out in order to settle the matter. The spot was duly observed (see Figure 2.5) and in an example of cruel irony or comic justice, it has ever since been known as Poisson's Bright Spot.

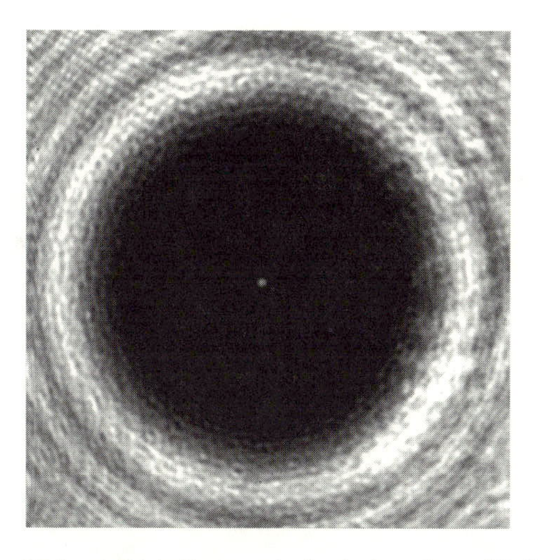

Figure 2.4: An image of Poisson's Bright Spot seen in the shadow of a small ball bearing.

Once more a property of light had been explained by adopting a wave theory. An overall explanation of the nature of light as a combination of oscillating magnetic and electric fields came from James Clerk Maxwell in 1873 (see Appendix B for more details), but by then the issue was pretty much resolved anyway. With an understanding of the wave nature of light, telescopes could now be designed for optimum performance. For one thing, it became apparent from theory that larger telescopes were better, and that images could not be magnified *ad infinitum* to reveal ever more detail. We will come back to this in the next section but for now we have reached the end of our history of telescopes.

From this point on, the design of telescopes changed in minor ways, but a discussion of this would become a listing of observatories with incrementally larger telescopes, which would be fairly tedious. Likewise, the emphasis of this book is the telescope itself, not the many historical discoveries which resulted from these constructions. However, we will return to discuss some key milestones in later chapters. But first we will look at just how telescopes work and at some of the issues involved in building them.

Chapter 3

How a telescope works

Imaging

Light can be focused by a lens using refraction. Refraction is the change in direction of light when it goes from one substance to another – e.g. from air to glass. An example of how an image is formed with a lens is shown in Figure 3.1. Light from a point on an object (in this case, the top of a candle flame) spreads out in all directions as indicated by the arrowed lines. That light which is collected by a lens is recombined to a particular image point. All points on the object are similarly recombined to give an overall image. The magnification is the image-lens distance divided by the object-lens distance (and is a demagnification in this case).

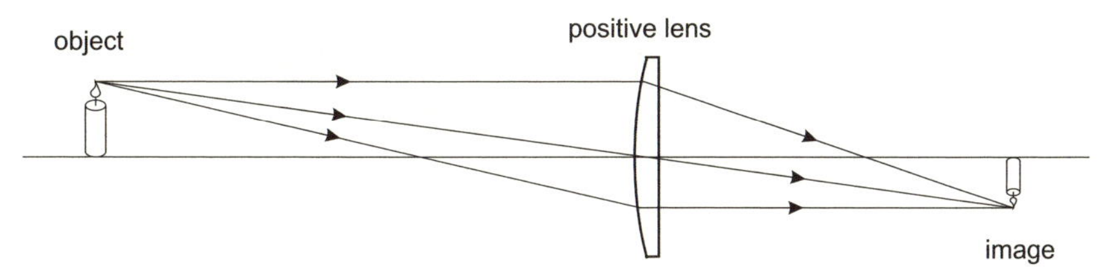

Figure 3.1: Imaging with a positive or convex lens. Rays of light (lines with arrows) from each object point are refracted to a corresponding image point.

For rays emanating from a particular point on an object, there is only one point at which they all come together on the image. If we place our screen inside or outside the image location, the image we get will be out of focus. In the case of the positive lens shown above, the image is a 'real' image. That is to say, we can physically touch it, or perhaps place a screen there and have the image projected onto it. In the case of a camera, we use film instead of a screen, and we can expose the film to permanently record the image. Your eye performs the same imaging procedure for objects in front of you. It may seem hard to believe at first, but all the images projected onto your retina are in fact upside down. The brain merely adapts to that fact and decodes the information in such a way as to 'tell' you how the world looks right side up.

Mathematically, we use a property of the lens called the focal length in order to determine the image location from a given object distance. The focal length is a measure of how much the lens will bend the light. More precisely, it is the distance at which rays from infinity are brought to a single point. If the 'object' in Figure 3.1 were placed a long, long distance away, the image would be at the focal distance. For the purposes of this book, all we need to know is that a longer focal length lens will bring rays together at a greater distance from the lens.

Refracting telescopes

When we put two positive lenses together we can make a telescope. In this case, the first lens (the primary) simply produces an intermediate image of the distant object, while the job of the second lens (the secondary) is to magnify the image. A diagram of how this works is given in Figure 3.2. Generally, the primary is a large lens with a long focal length, while the secondary has a short focal length to give us a large amount of magnification. In a simple telescope such as this one, the primary was traditionally called the 'objective' and the secondary the 'eyepiece'. In more complex systems these terms have been abandoned as there are many more elements and they don't serve their original purposes any more (for example, we no longer use our eyes for astronomical work).

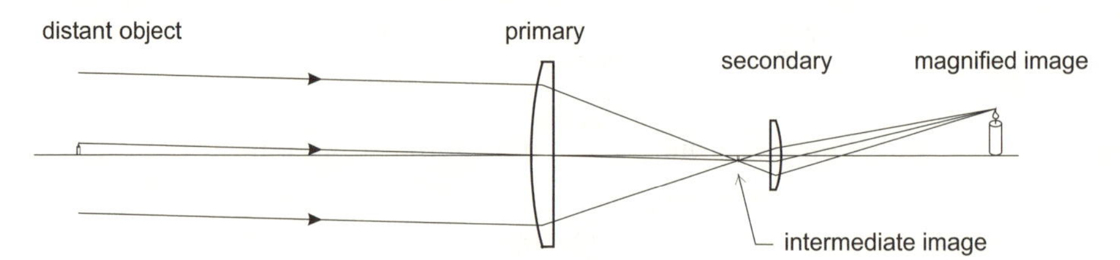

Figure 3.2: A telescope. The combination of two lenses allows us to magnify the image of a distant object (a candle).

To see how the telescope magnifies, take a close look at the size of the object and the final image. The object makes a small angle to the primary lens, while the image subtends a large angle to the secondary. In simple telescopes such as this one, we can talk about magnification in terms of the amount by which the angle is magnified. That is all that the lenses are doing – making every angle within our field of view much larger. Imagine a tree in the distance which is brought close to your eye. At a distance, the entire tree makes a small angle, while up close even a branch may make an appreciable angle. Angular magnification serves as an acceptable method for describing how powerful a telescope is in the case of a simple device like that shown here, but is not the way astronomers do things when dealing with mountain-top devices. Also, as we will see later on, magnification is often abused by telescope makers in trying to get people to spend unnecessarily large amounts of money on the wrong instrument.

The telescope shown in Figure 3.2 is sometimes called the 'astronomical telescope'. As drawn here, it is the simplest version to understand, and it produces a real image which can be captured on film. If we move the secondary a little closer to the primary, the direction of the rays will change dramatically, resulting in a virtual image, as shown in Figure 3.3. A virtual image is one which can be seen, but cannot be projected onto a screen – we cannot interact with a virtual image. An example is the reflection of our world in an ordinary bathroom mirror – we can see objects, but we can't physically touch them, and we can't project that image onto a screen. In fact, the reason we 'see' this image at all is because our eye has another lens which turns the virtual image into a real one on our retina. Virtual images are okay for a telescope which we want to look through, but are not suitable for recording images. Another problem with the astronomical telescope is that the image is inverted (upside down), so it can be a little confusing to look at scenes through such a telescope.

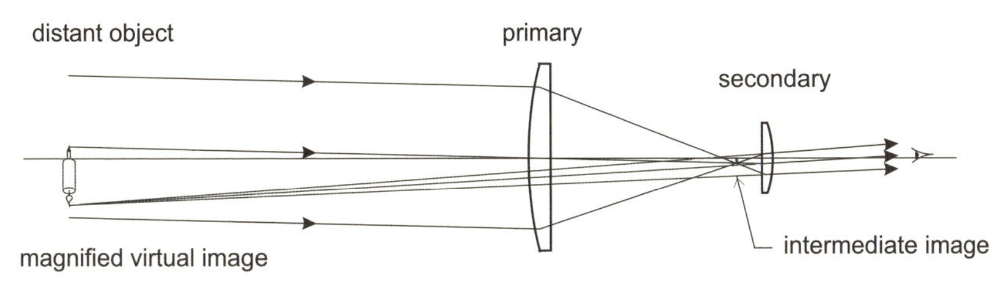

Figure 3.3: The astronomical telescope for viewing with the eye. By looking back through the telescope, we can see a magnified virtual image of the object inverted in the distance (shown dashed).

Although this is a little harder to understand, the principle is still the same – the primary produces an image, and the secondary magnifies it. This is the underlying principle behind all telescopes, even though the intermediate image may not be easily located. Because making a telescope uses this rather simple method, there are many different variations on the design. One of these alternative configurations is actually the one which was discovered first – the Galilean telescope. To understand its operation, we will need to introduce the negative lens. This concave or divergent lens will cause parallel rays to diverge instead of converge as they do for a positive lens. Negative lenses generally produce virtual images, as we can be seen in Figure 3.4.

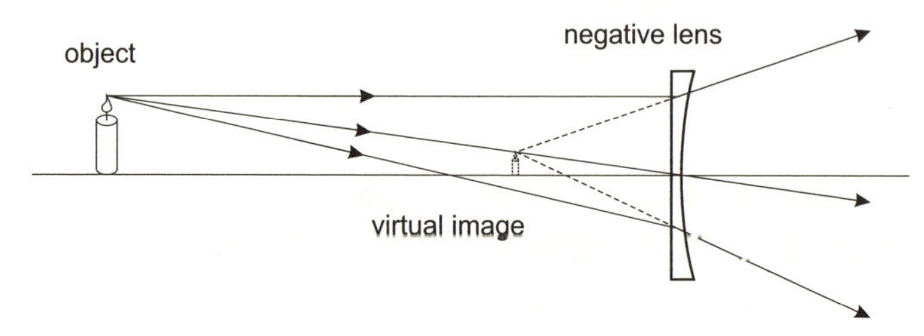

Figure 3.4: Imaging with a negative or concave lens. Rays of light from each object point are refracted in such a way as to diverge. Looking back through the lens, we would see a virtual image where the ray paths converge (shown dashed).

The Galilean telescope still has a convex lens as the primary. The difference is that the negative secondary lens is placed before the intermediate image is formed (as shown in Figure 3.5). If we look through the telescope, we will gather the divergent rays and see the virtual image of the object lying a long distance away, but greatly magnified. The ray-tracing in this image seems a little convoluted, and this is why I began the description of how a telescope operates with the simpler astronomical version.

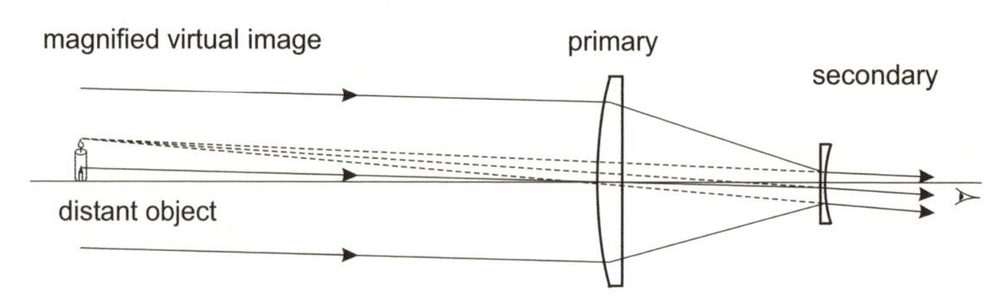

Figure 3.5: A Galilean telescope consists of a positive primary and negative secondary. Following the path of the rays shows that we end up with a greatly magnified virtual image an infinite distance away. Rays are refocused in our eye to produce a real image.

The Galilean telescope is shorter than its sibling, and has the added benefit that objects seen through it appear the same way up as they are in reality. This telescope was the first 'spyglass'-type telescope used by mariners. It is simple and relatively easy to make. Of course, in professional telescopes, lenses are rarely used. Instead we achieve the same imaging properties using concave and convex mirrors. The reason for this will be explained in detail later, but one important reason is that it reduces the overall length of the telescope.

Reflecting telescopes

Since light bounces back and forth over the same area in reflective telescopes, things get very confusing if we draw rays as we did before. So at this point we will simplify the drawing so that we don't confuse the issue with lines everywhere. Since the basic operation is still similar to that of the two refractor telescopes described above, you can always return to these descriptions to see how it would work. The incoming light from a distant object is now shown as parallel lines entering the telescope.

The simplest reflective telescope is the Newtonian telescope, named after its inventor, Sir Isaac Newton. Instead of a convex lens, the primary is a concave mirror, which forms an image back in the direction of the incoming light (Figure 3.6). Before the rays reach this focus, a small flat mirror is used to reflect this image light out to the side. This is so we don't go sticking our heads in front of the mirror, and in doing so block out the light from the object that we are trying to see. A small secondary lens is then placed beyond the focus to magnify the image. Upon careful examination, you may notice that this telescope is essentially the astronomical telescope, but with a primary mirror instead of a lens, and the secondary lens unchanged. The primary mirror in this case has to be parabolic (not spherical as Newton originally thought) to avoid blurring errors (which we will discuss later on).

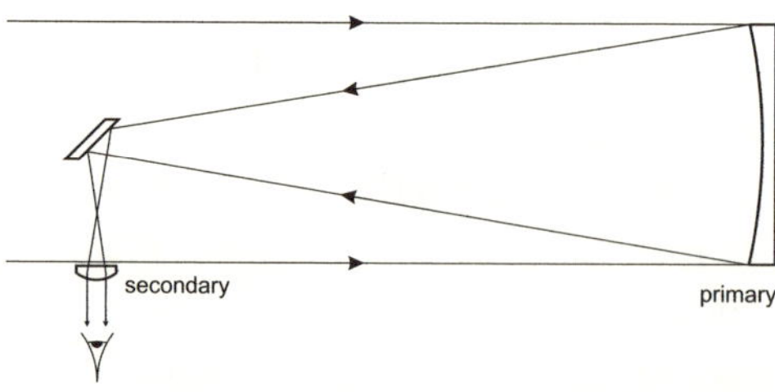

Figure 3.6: A Newtonian telescope consists of a parabolic primary mirror which focuses light onto a pick-off mirror, thus directing the intermediate image to the side. This image is magnified using an eyepiece lens.

The next basic telescope type is a reflective version of the Galilean telescope that incorporates a concave (positive) primary and a convex (negative) secondary. In this case, the light is focused by the first mirror onto the convex secondary. This mirror reflects the light back through a hole in the primary, to create a real image (Figure 3.7). By folding the light back on itself, the length of the overall instrument is greatly reduced. Those of you paying attention may notice I slipped a little change in here – the negative secondary is forming a real image, unlike the virtual image produced in the lens version shown in Figure 3.5. The truth is that the secondary is making a real image, but I have specifically taken this mirror to be a weakly curved mirror for this purpose. If it were replaced with a highly curved mirror, we would get precisely the same result as for the Galilean telescope.

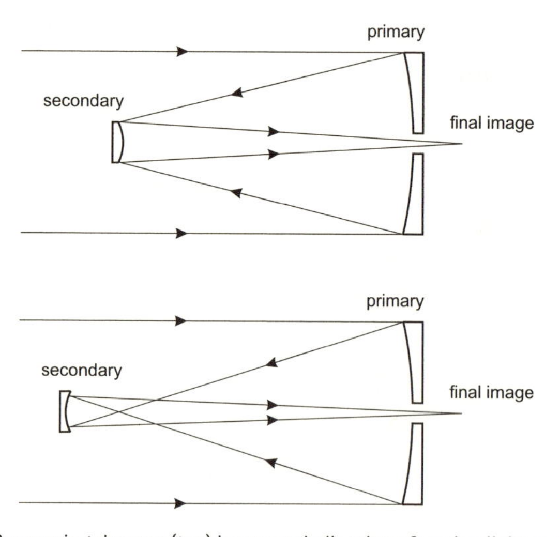

Figure 3.7: The Cassegrain telescope (top) has a parabolic mirror focusing light onto a convex secondary which reflects the light back through the primary. The Gregorian telescope (bottom) has a concave secondary.

The top reflector telescope shown in Figure 3.7 is called a Cassegrain telescope after the French priest Laurent Cassegrain who made the first working model. With very minor modifications, this basic design is the one most commonly used in amateur and professional astronomy. It is compact and images along the line of sight (instead of at right angles). The second reflector shown in Figure 3.7 is the Gregorian telescope named after its inventor, a Scottish professor James Gregory.[10] This design is rarely used since it has few significant advantages over the Cassegrain configuration, but is longer – which makes it more cumbersome to steer and requires a larger dome for it to sit in.

Let's take a short moment to address a common question: can you turn a telescope into a microscope by looking through the wrong end? The answer is: not really. Telescopes and microscopes are very similar optical instruments, and work on the same principles to produce magnified images of objects. In the simplest version, they both have two lenses – one of which is an eyepiece designed to magnify an image into our eyes (or onto a recording medium). However, both produce an increase in magnification in one way only. In most cases, if you turn them around you will only produce a demagnification, not a magnification.

The telescopes discussed here are all variations on the two basic types: the astronomical and the Galilean. There are, however, many further possible configurations. Some of them have many more than two mirrors, while others have weird combinations of concave, convex and flat mirrors. Many of them have specialist applications and others have been suggested by optical designers but never actually adopted as they are impractical. With every new observatory created, much effort goes into the configuration of the optics, and many designs are evaluated for their various benefits and trade-offs. We will discuss some of these issues later, but it is important to remember that while the optical design may get fairly convoluted at times, the basic principles described here still apply: the large primary element is used to produce an image, and the secondary element provides magnification of the image. It may seem that by using a secondary which provides increasingly higher magnification, we could get the telescope to produce unlimited magnification. In fact this is not the case, and the reason comes down to the wave nature of light. To understand this better, we need to step back a little and look at some of the basics.

Chapter 4

The perfect telescope

Diffraction and the perfect image

Much like a water wave, a beam of light has an advancing wavefront which will interact with matter. Imagine a water wave passing by a reef, then look at the two diagrams below (Figure 4.1). The first case just looks wrong, and the second seems right, but why? In the first case, the wave is being broken apart by the reef, whereas in the second case, part of the wavefront bends around the barrier. This bending of a wave when it encounters an obstacle is called diffraction. Since light is a wave, it does the same thing. If we are to describe truly the way light behaves, we must take into account the fact that every time we restrict a wavefront of light by putting an aperture in front of a beam, we will find this diffraction occurring. As we will see, diffraction has some implications on the images produced in telescopes.

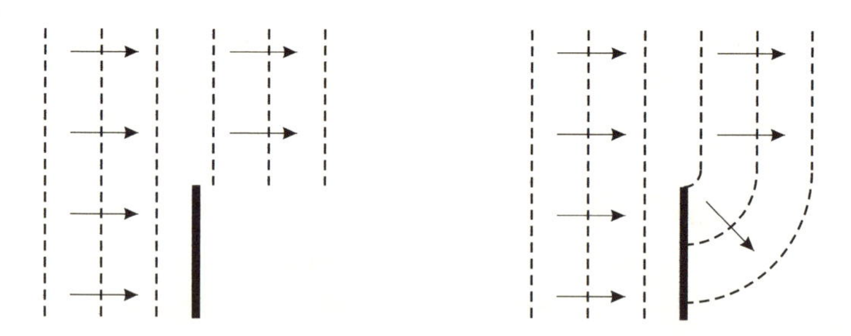

Figure 4.1: Diffraction. An advancing wavefront (dashed) is shown meeting a barrier. The left side shows the case with no diffraction and on the right the wavefront diffracts around the barrier as we are used to.

To begin with, let's return to the simple diagram we have of a telescope, as shown in Figure 3.7. Notice how we have used straight lines to indicate how the light is focused to an infinitely small point. Diffraction means that, for the most part, this idea of light travelling in straight lines is true, but in fact the first element (the primary) acts as an aperture which produces some bending of the light at the edges. In doing the mathematics (which I will spare you), we find out that diffraction will result in a focal spot which is not infinitely small but actually has some size. In fact, the spot produced when focusing a point source of light through a circular lens looks like that shown in Figure 4.2 – a fuzzy spot surrounded by faint rings. *This is actually what a star looks like seen through a perfect telescope in space*, and it has huge implications on how much we can magnify images.

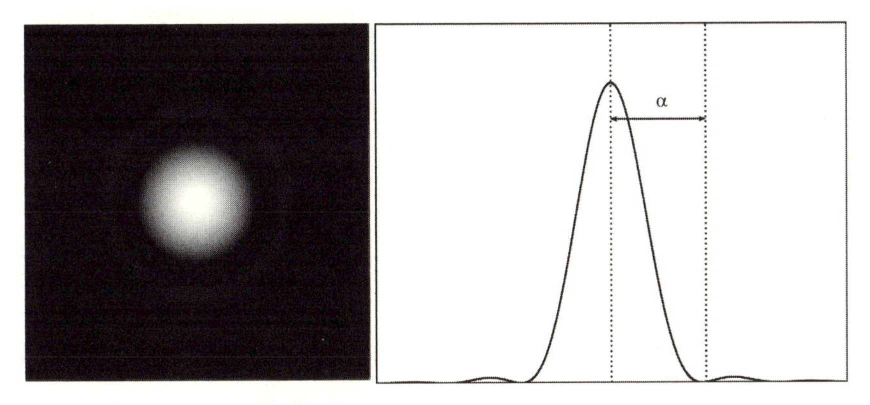

Figure 4.2: An Airy spot (left) is what a star will look like when viewed through a perfect telescope with a circular primary. It consists of a bright spot surrounded by dim circular rings. On the right is a plot of the intensity varying on a slice through the centre. The angular distance from the maximum to the first minimum (as shown) is α.

The spot you see is called an Airy spot – named after the Englishman, George Airy, who first described it. If we take a slice through this image, we can see how the intensity of the spot varies, as shown on the right of Figure 4.2. We can now see a large peak (the bright spot), going to nothing (the first dark ring surrounding the central spot) on either side. The angular distance from the maximum to the first minimum (as shown) we will give the symbol α. We can write a formula for calculating the value for the angular resolution (α) in terms of the wavelength of light (λ), and the diameter of the primary (D) that we are using to focus the light:

$$\alpha = 1.22\frac{\lambda}{D}$$

If you are still with me up to here, this is where the pay-off begins. I had to promise the publishers that there would be only one formula in this book, so you can probably understand that it must be an important one. What this formula tells us is the size of the Airy spot we get when we focus light of a certain wavelength down with a lens (or mirror) of a certain size. Remember that the spot represents the image of what a point-like object (a distant star, for example) looks like in our telescope. The spot is very tiny, often a fraction the width of a human hair. Extremely small, but still finite. We can also see from this equation that the bigger the diameter of the aperture, the smaller the Airy spot produced will be. The importance of this will become obvious below.

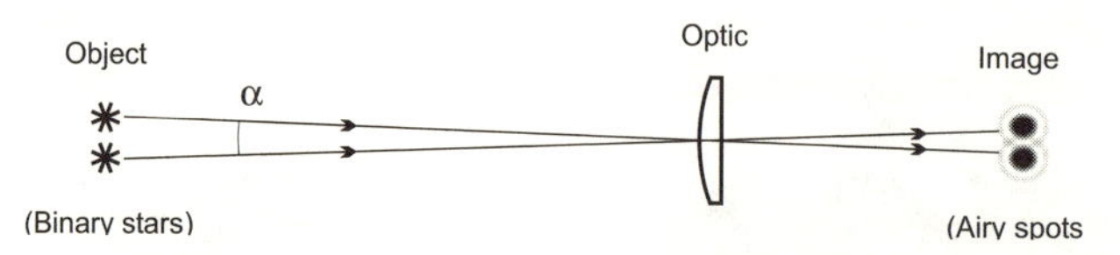

Figure 4.3: Binary stars will be imaged into Airy spots. The size of the spots is determined by the diameter (D) of the optic and the wavelength of light used. We can resolve the two stars if the angle is larger than α.

Resolution limit

If we try to magnify this image of our star in an attempt to see more detail, we will find that this only makes this Airy spot bigger – it doesn't let us see anything new. So this represents some limit on how much detail we can see. To understand this better, let's consider the case of a binary star system (a pair of stars orbiting each other) as seen through our telescope. If the two stars are far enough apart, our image will consist of two Airy spots next to each other, as shown in Figure 4.4(a). Now let's look at two stars that are a little closer together, as shown in Figure 4.4(b). In this case, they are so close that the Airy spots overlap, but the final image does still appear to be two spots. If we continue to decrease the separation of the stars we might get to a point where the stars overlap, as shown in Figure 4.4(c). At some point we have to draw the line and say, 'Yes, I'm sure there are two stars', or 'No, I can't tell if there are two stars or just one'. This arbitrary point is called the *resolution limit*. In the nineteenth century an Englishman, Lord Rayleigh, decided on a mathematical description of where to set this limit. He stated that the simplest solution was to say that it was where the maximum of one spot sat right over the minimum of the other. That is, when the angle between the stars is equal to that given in the formula above.

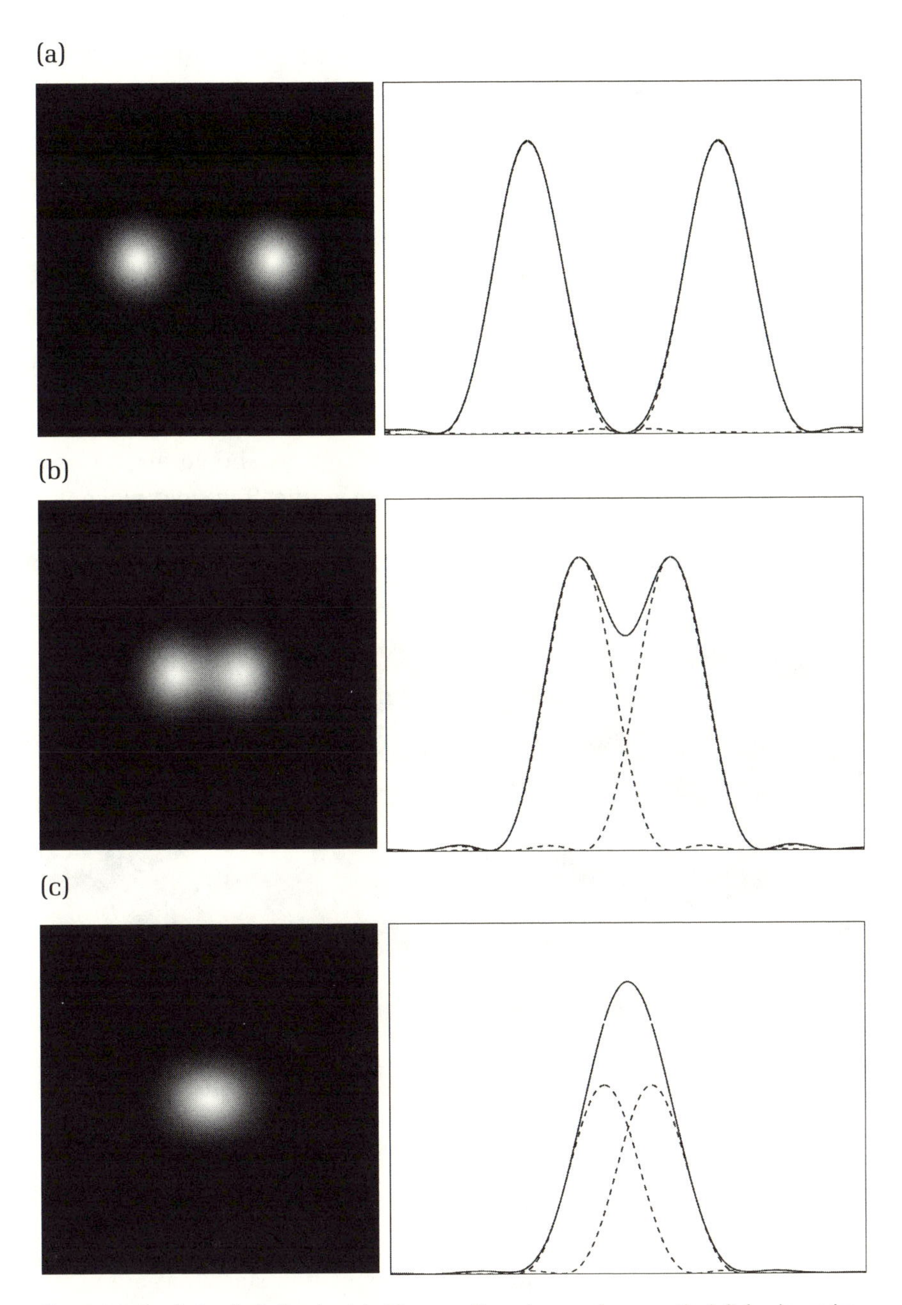

Figure 4.4: Resolution limit. The simulated images of two stars are shown on the left for decreasing separation going from **(a)** to **(c)**. On the right is a plot of the intensity for a horizontal slice through the centre of each image. The dashed line is the intensity of each individual Airy spot, and the darker line represents the combined intensities.

Another way to think of this resolution limit is to realise that every point on our object will be blurred into the Airy spot after it passes through our telescope. The amount of blurring depends on the size of the primary element – larger primaries give less blurring. So in order to improve the clarity of our images (i.e. increase the resolution), we need a larger collecting element. Again, it must be stressed that magnification doesn't come into it. The blurring has been provided by the first element in our telescope, and no amount of magnifying will make things clearer – they will just be bigger. When buying a telescope, you can often tell the difference between those aimed at the recreational astronomers, and those for serious amateur astronomers. Often the cheaper telescopes will feature a note on the box advertising '50X magnification!' and other such claims. Magnification is important, but only up to a point. Take a look at the images in Figure 4.5, where a binary star seen through the same telescope is shown at low and high magnification. Notice that the extra magnification does make the image bigger. However, since the stars were not resolved before, they will not be resolved with increased magnification.

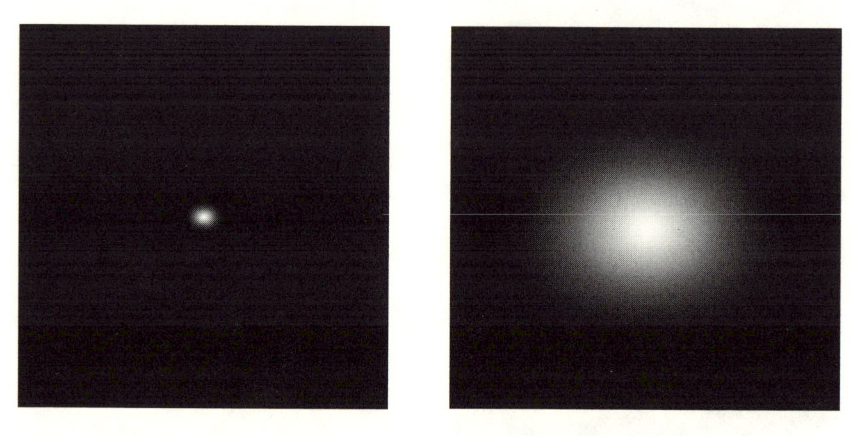

Figure 4.5: The same binary is seen through the same telescope at low (left) and high (right) magnification. The unresolved stars are still unresolved even with the higher magnification.

So what does this all mean? Well for one thing, it means that for any optical element (or combination of elements) there is a limit to how much detail we can see. From the equation above, we can see that the only way we can separate the images of two stars that are close together is to make α smaller, by:

a. choosing a shorter wavelength of light
b. using a larger diameter primary.

If we are observing at some set wavelength (in the visible, say), then that really only leaves us with the second option. This is one reason why we want bigger and bigger telescopes – the larger the diameter (D), the better the resolution. To see this, we can look at the images in Figure 4.6, below, where the images of the **same binary** are seen through a small and a large telescope. The stars are not resolved by the small telescope, but the larger one has smaller Airy spots, so the two stars are clearly seen as two separate objects.

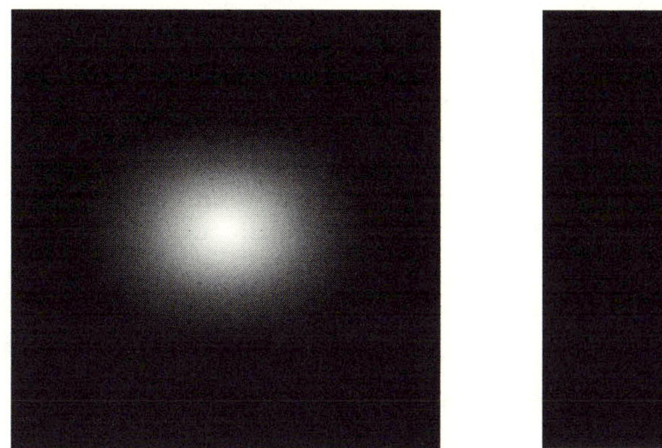

Figure 4.6: Simulated images of the same binary as seen through a small telescope (left) and one five times larger (right) show how increasing the diameter of the primary improves resolution.

So let's try some numbers by using the 2.4-m diameter Hubble Space Telescope (HST) as an example. While it images over a wide range of wavelengths, the best resolution will come from the shortest operating wavelength of $\lambda = 110$ nm (ultraviolet). This means that the resolution limit for the HST is $\alpha = 56$ nanoradians. This is a very small angle indeed – equivalent to 3 millionths of a degree, if that helps at all. So this is the smallest angle the Hubble can resolve – but what does this mean in distance? Well, say we were using the HST to observe the Moon, which is roughly 380,000 km away. This resolution limit means we could resolve two points separated by 21 m. So if we were to use the Hubble to take photos of the Apollo lunar modules, we would just see a single bright point of light. However, if the lunar rover were parked over 21 m from the lander, we might be able to pick up two separate bright spots of light under the right lighting conditions.

The resolution limit is somewhat arbitrary in that it was chosen rather than actually being a firmly calculated value. In 1916, Englishman Charles Sparrow decided that the Rayleigh criterion was a little too large. He said that two stars could be discerned as such if they were separated by an angle of just $1.0 \times \lambda/D$. Not a huge improvement, and since this was an equally arbitrary definition, it did not supplant the Rayleigh value of 1.22. It did, however, get his name in the books with the 'Sparrow limit' even if it is rarely used, so for this scientist at least it did serve some purpose.

I mention this merely as a way of emphasising that while the Rayleigh limit is as good a way as any other of defining the resolution limit of a telescope, it is not a rigid limitation. Unlike the speed of light, which is a true limit, this is more like a posted speed limit on the roads – you might be able to beat it, but you'd better be very careful and know exactly what you're doing. As a side note, while Lord Rayleigh is best remembered for this optical definition, he was actually given a Nobel Prize in 1905 for 'his investigations of the densities of the most important gases and for his discovery of argon in connection with these studies'. As far as I am aware, his work in the field of optical resolution has proved far more important in the long run.

When it comes to imagery, resolution makes a difference in how much detail we see. Nowhere is this better demonstrated than in observations of Mars. In 1877, Giovanni Schiaparelli described in great detail the 'canali' or 'channels' that he saw on the planet's surface. Percival Lowell took this further, producing elaborate maps as proof of intelligent beings on Mars who built these canals to move water from the poles to lower latitudes. Later studies of Mars with better telescopes and improved resolution showed that the 'observed' channels were in fact not there. Many explanations from physiological to merely wishful thinking have been used to explain these 'observations'.[11]

A more notable example of the dangers of observations made with insufficient resolution came in 1976 with the *Viking* missions to Mars. NASA launched two probes which each consisted of an orbiter and a lander. The *Viking Orbiters* were designed to take high-resolution images of the planet's surface from an elliptical orbit using a telescopic lens with a diameter of 0.2 m. One of the images was of a formation which looked eerily like a human face. NASA scientists were quick to point out that this image was merely a freak combination of lighting on a completely random rock formation. This is compounded by the fact that our brains have a penchant for forced pattern recognition.

Naturally, none of this carried any weight with those who saw it as proof of extraterrestrials. The matter of the 'Face on Mars' remained irresolvable as there was simply no way to improve the resolution of the image. In 2001, however, NASA parked the *Mars Global Surveyor (MGS)* in orbit around Mars. With a larger, 0.4-m diameter telescope, a much higher-resolution image was taken.[12] In Figure 4.7, we can see a comparison between the *Viking* and the *MGS* images. The improved image clearly shows that the 'face' is in fact an artifact of lack of resolution, but some 'facial' features are still imaginable. In fact, it is possible to turn the higher-resolution image back into a face by squinting at it.

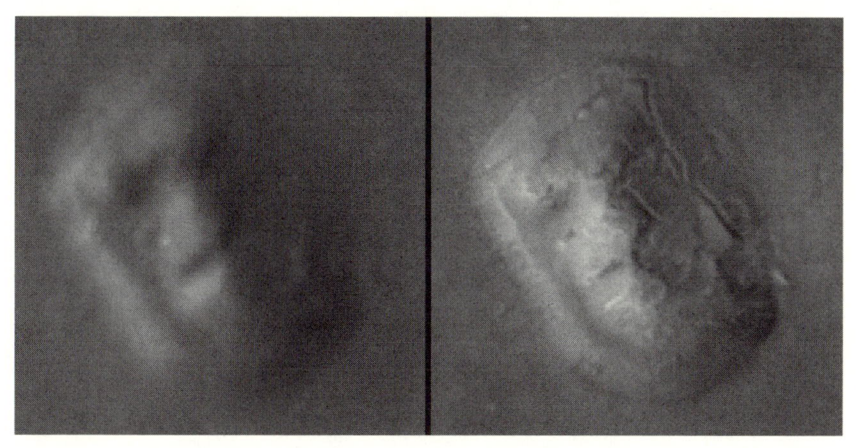

Figure 4.7: The 'Face' on Mars revealed. On the left is an image of a surface feature on Mars taken by the *Viking Orbiter* in 1976. A higher-resolution image taken by the *Mars Global Surveyor* probe in 2001 (right) reveals more details with less anthropomorphism. Courtesy: NASA/JPL/Malin Space Systems.

The important points to remember are that we can't magnify images indefinitely, and that resolution can only be improved by using shorter wavelengths of light and bigger telescope primaries. Already you can see that this suggests that there is something behind the 'mine is bigger than yours' bluster that astronomers think about when choosing a telescope to use for observations. However, there is more to it than that – a larger mirror will collect a greater amount of light, and thus give us brighter images of distant objects and allow us to take images in a shorter amount of time.

There are two final details to note before we finish this discussion on resolution. The formula on page 45 really only applies to two point sources of equal brightness. If one of the stars is much brighter than the other, we might have trouble seeing a small Airy spot in the halo surrounding the bright Airy spot. In this case, the formula does not quite work. The significance of this will be addressed later on when we consider the problems of seeing planets orbiting distant stars. The other important issue that the resolution formula assumes is that the telescope being used is of good quality. Just like our eyes, telescopes have all sorts of problems which can cause blurring. If such aberrations are present, then the resolution formula can no longer be applied and all bets are off as to what the resolution limit is.

Chapter 5

When good telescopes go bad

Aberrations

You may have not noticed it, but throughout the previous chapter, I was very careful to talk about 'perfect wavefronts' and 'perfect optics'. By this I meant that in order to achieve the best possible result of an Airy spot, we must have an incoming wavefront that is perfectly flat and optics that have perfectly shaped surfaces (e.g. a perfect parabolic mirror in the case of a Newtonian reflector). A star is essentially a point source of light an infinite distance away. Light coming from a star travels through a vacuum, and as such serves as an ideal, perfectly flat wavefront. But just how perfect is 'perfect'? When Lord Rayleigh formulated his resolution criterion, he realised that the presence of distortions in the input wave would have an effect on the Airy spot, and this would in turn have an effect on resolution. We have come across this in our everyday lives when trying to look through distorting shower glass. Much to the disappointment of potential voyeurs, our ability to form images of objects on the other side of the distorting glass is affected by the presence of an imperfect (though transparent) medium.

With this in mind, Rayleigh was careful to specify that so long as the wavefront was never distorted by more than a quarter of the wavelength of the light, we could still consider the Airy spot 'perfect'. If the distortions in the final focused wavefront get worse than this, then all bets are off, and the Rayleigh limit no longer applies. In the case of a perfectly smooth wavefront, we would say we have 'diffraction-limited imaging' – a fancy phrase meaning that the only thing that limits the resolution of the telescope will be the inherent wave-nature of light. However, if we have a situation where the advancing wavefront gets distorted beyond this, the Airy spot starts to degrade into a larger distorted blob. Since the size of the blob determines resolution, a degraded spot will mean reduced resolution.

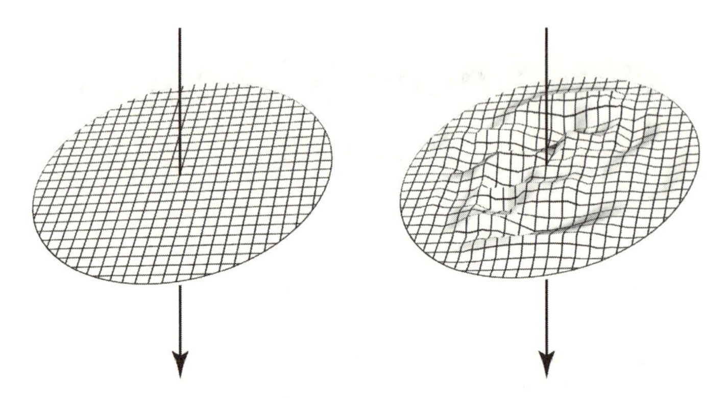

Figure 5.1: This image shows what we mean by wavefront distortion. The wavefront on the left has an ideal flat phase, while the one on the right is a more typical wavefront entering our telescope. The Rayleigh criterion says that the wavefront should be flat or spherical to within a quarter of the wavelength of light for diffraction-limited imaging.

If we focus a perfect wavefront with a perfect telescope we will get the Airy spot we desire. However, in practice, there are many issues which can prevent this from being the case. These include having a distorting medium between the telescope and the star – such as the Earth's atmosphere. Another source of wavefront errors could be the telescope itself; it could be constructed from poor-quality optics, high-quality optics which are misaligned, or the incorrect type of optics. Each of these potential problems can keep telescope makers awake at night (an ideal state of affairs for astronomers, but not for optical engineers). Most of the work involved in telescope construction is spent trying to nurse a wavefront of light unimpaired all the way to the image plane. A multitude of alignment, construction and other design issues need to be considered to ensure that not even the minutest of errors is introduced. The purpose of this chapter is to give a feel for the effort involved in creating a high-quality magnified image.

To begin with, we will consider a telescope located in space where the light is unaffected by the atmosphere all the way to the telescope. Some potential problems with the telescope are that the optics could have less than perfect surfaces, be the wrong shape, or perhaps be misaligned. Any of these problems will give rise to imperfect focusing. Let us consider the first case: an imperfectly made surface. A parabolic mirror will focus starlight to an Airy spot, so long as the surface of the parabola does not have severe distortions. Once again, we require the distortions introduced to be less than a quarter of a wavelength of light, often written simply as $\lambda/4$. In practice, this would limit the largest bump on the mirror surface to be less than $\lambda/8$ (as the light will travel twice the distance in reflection).[13]

The world's largest telescopes have mirrors of around 10 m in diameter. If such a mirror were expanded to the size of Australia, then, according to the Rayleigh criterion, the largest 'mountain' or 'valley' on the surface would have to be no higher than a golf ball! This is impressive enough, but in a typical telescope light is bounced off several mirrors before it finally arrives at the camera. As a worst case then, we have to assume that the aberrations are compounded each time, so the overall requirement on each optic may have to be increased to $\lambda/20$ or better. It is for this reason that the requirement of the Hubble Space Telescope mirror was a surface good to $\lambda/60$. Such extremely well-figured and polished surfaces require extraordinary manufacturing techniques, which will be discussed in a later chapter.

Lenses have different problems. Unlike mirrors, they have two surfaces instead of one which have to be shaped and polished, but in fact the requirements on the surfaces are slightly more relaxed. Refraction bends the light very gradually, and as such the light is less sensitive to errors in the surface. The nature of reflection means that the effect of even the smallest bump is magnified greatly. Typically, the surface of a lens can have a surface error around six times greater than that permissible on a mirror and nonetheless give diffraction limited imaging. To understand this, think about a pool of water. There is no problem in seeing objects at the bottom of the pool, even when the surface is disturbed. Conversely, scenes reflected in the surface will be distorted significantly by the slightest of ripples. This is one of the reasons why reflecting telescopes were slow to become popular – they were much more difficult to make. It also explains why Galileo was able to make high-quality lenses even without modern technology as a guide.

But refractors are far from the ideal type of telescope to build for several reasons. Firstly, there is the fact that the glass has to be very homogeneous. That is to say, we have to make sure that there are no bubbles or cracks inside, and that the density is extremely even throughout, so that all parts of the glass bend (refract) the light by the precisely desired amount. All well and good, but even with the ideal piece of glass there is an even bigger problem called dispersion. In 1704, Newton showed how a prism of glass can break up light into its component colours. This is called dispersion and is a result of the fact that different wavelengths of light bend at different angles as they go from one medium to another (such as from air to glass or from glass to air). If you take a look at a lens side-on, you can see that the edge does look a little like a prism, and, sure enough, we will get dispersion occurring when we focus light with a lens. The result of this 'chromatic aberration' is that the lens will have different focal lengths for different wavelengths. Such a refractor will only be in focus at one particular wavelength, and our images of stars will have colourful blurry halos.

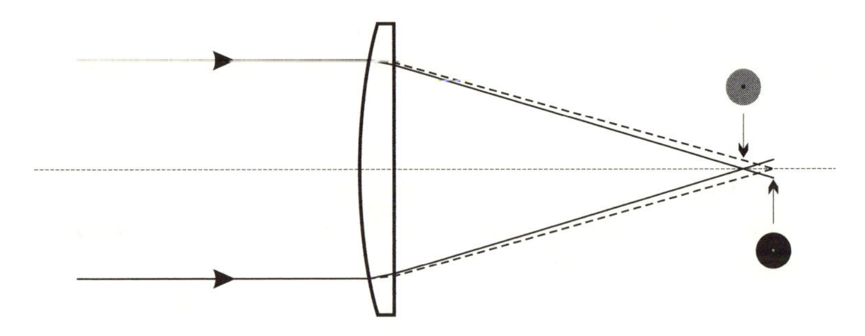

Figure 5.2: Chromatic aberration. Because different colours bend at different angles, they focus at slightly different distances. The bluer wavelengths (solid line) focus at shorter distances than redder wavelengths (dashed). The result is a sharp focus at one colour surrounded by a blurred disk of another.

To reduce this effect we can make our lens out of a combination of two or more materials, each one designed to minimise the dispersion. This type of lens is known as an achromat, and is shown in Figure 5.3. Achromatic lenses are more expensive than single component lenses (called singlets), but are relatively easy to make for the small refractors available in stores. However, as the diameter increases, the costs rapidly escalate. This is one reason why the world's largest refractor, the Yerkes telescope built in 1897, is a mere one metre in diameter.

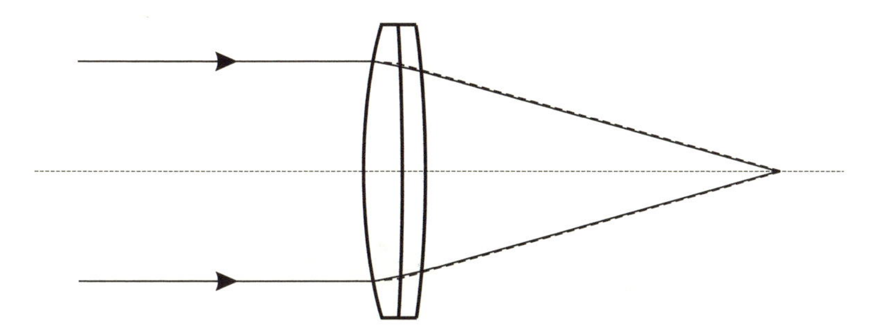

Figure 5.3: Achromatic lens. Making the lens from two or more different types of glass can recombine the colours and reduce chromatic aberration.

Unlike refraction, there is no dispersion on reflection, so all colours are focused in the same location. This is probably the main reason for favouring reflectors over refractors, but there are others. To begin with, mirrors have just one surface which needs to be polished to a high degree of precision, as opposed to two (or more in the case of achromatic lenses), thus simplifying the fabrication process. Reflector telescopes tend to be shorter as a result of the folded paths that the light takes, so they take up less space and permit us to use smaller domes to house them in. Lastly, mirrors are easier to support. Whereas a lens has to be held around the edge, which puts lots of strain on the glass, the large primary in a reflector can be supported at many points around the edge and rear. Better support means less chance that our carefully constructed mirror will bend and distort beyond the quarter wave(length) limit. For these reasons, all of the world's major telescopes for the last century have been reflectors.

Even with mirrors polished to a perfectly smooth surface devoid of any bumps, we may not produce diffraction-limited images. This is because the mirror may have the wrong overall shape. Until now, I have restricted the discussion to parabolic mirrors since only a parabolic surface can take a plane wave of light to a single focal point. This makes it the required shape for the Newtonian reflector. However, mirrors can be made in many other curves (called conics) such as spheres, hyperbolas and ellipses. Take, for example, a spherical mirror, which has more curvature than a parabola, as shown in Figure 5.4. If a spherical mirror is used in a Newtonian telescope instead of a parabolic one, the outermost rays will focus closer to the mirror than the inner rays, producing an error called spherical aberration, as shown in Figure 5.5. Instead of a perfect focus, the rays then come to a blurred spot with a size significantly larger than that predicted by diffraction.

In fact, spherical aberration is precisely the problem that Rayleigh was thinking of when he coined his quarter wave criterion.

Figure 5.4: The paraboloid is shown solid, the sphere is dashed. Notice the increasing separation between the two conics as the diameter increases.

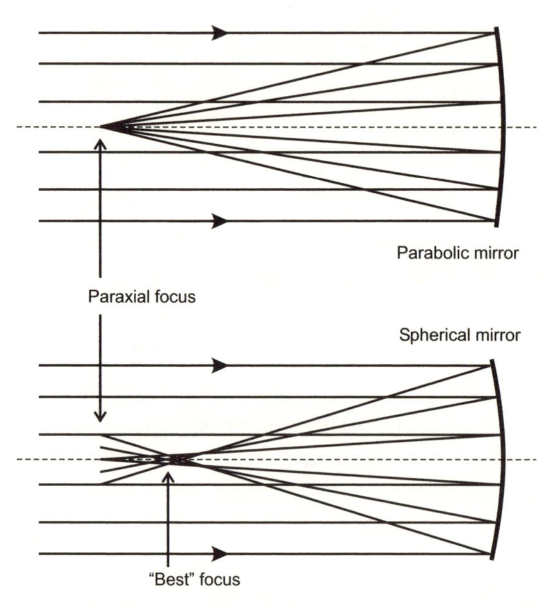

Figure 5.5: Parabolic mirrors reflect all distant rays to a common focus. With spherical mirrors, the rays towards the edge are reflected to focus short of the true, 'paraxial' focus.

In reflection then, we cannot permit the mirror to have a shape that deviates from a parabola by more than an eighth of a wave. During the fabrication process, simple grinding forms a spherical shape which then has to be further ground down towards the edges to give a parabola. In Figure 5.4, we can see that the larger the mirror gets, the greater the difference between a parabola and a sphere. So, while it is relatively easy to make a parabolic shape for small mirrors, this becomes increasingly difficult as the diameter is increased – a big problem in the case of big telescopes.

All this would seem to indicate that only parabolic mirrors can be used in telescopes. In fact, this is not the case. It is true that spherical aberration would be a problem if any conic other than a parabola telescope were used as the primary in the single-mirror Newtonian telescope. However, when the telescope has multiple mirrors, we can use the later mirrors to remove spherical aberration introduced by the primary. Different conics introduce spherical aberration with a different sign, so they can be used to compensate for each other. For example, while a Cassegrain telescope uses a parabolic primary, a more useful version actually uses a hyperbolic primary and secondary – a modification known as a Richey-Chrétien telescope. This only works, however, if each of the hyperbolic mirrors is fabricated according to the precisely required shapes.

Spherical aberration is an example of a geometrical aberration – so-called because it is caused by an incorrect geometry in the telescope construction. Astigmatism is another common geometrical aberration. This is an error which causes the mirror to have a different focal length in one plane compared with that in a perpendicular plane. That is a convoluted definition, but it is easy to understand – just imagine taking a thin, curved mirror and pushing down on the top, squashing it slightly. What this does is change the mirror curvature in the vertical direction, therefore shortening the focal length. In the horizontal direction, the focal length remains unchanged. For people with astigmatic vision, distant point-like streetlights get distorted into oval shapes. In telescopes, the same thing happens, unless the aberration is really extreme at which point stars begin to look like plus-signs or lines, as shown in Figure 5.6.

Figure 5.6: Astigmatism. Images are shown scanning through a focus with a large amount of astigmatism present.

A second geometrical aberration is coma. It is harder to describe this one in a conceptual way, but the presence of this aberration will make stars look like comets – hence the name. The effect of increasing coma on the image of a star is shown in Figure 5.7. Coma is a big problem with parabolic mirrors, so it mainly appears in Newtonian telescopes. Both astigmatism and coma are so-called off-axis aberrations. That is to say, when we are looking at a star in the centre of our field of view (on-axis), it may be that our telescope has been well constructed and gives a perfect Airy spot focus. For a star at some angle off to one side, however, light reflects off the mirror at an angle. The result off this off-axis angle is to introduce astigmatism and coma.

Figure 5.7: Coma. Increasing amounts of coma from left (none) to right are shown affecting the focus.

Another pair of important aberrations is distortion and curvature of field. Distortion results in the position and shape of objects in our field of view becoming altered in the final image. Take a look at Figure 5.8, and you can see the effect of pincushion and barrel distortion on an image of a grid of straight lines. Most applications, such as photometry, spectrometry or polarimetry, can tolerate a large amount of distortion with little noticeable effect on measurements. For simply taking images, small amounts can be corrected so long as the nature and magnitude are well characterised. In fact, most camera lenses suffer from small amounts of this aberration and we barely notice it (for example, in the unusual case of Plate 4). On the other hand, we wish to avoid curvature of field whenever possible. This aberration has the effect of taking a flat image plane and making it curved. The result is that when using a flat recording medium such as film or CCDs (charge-coupled devices; see Chapter 6), the image is in focus in the centre but out of focus and blurred around the edges.

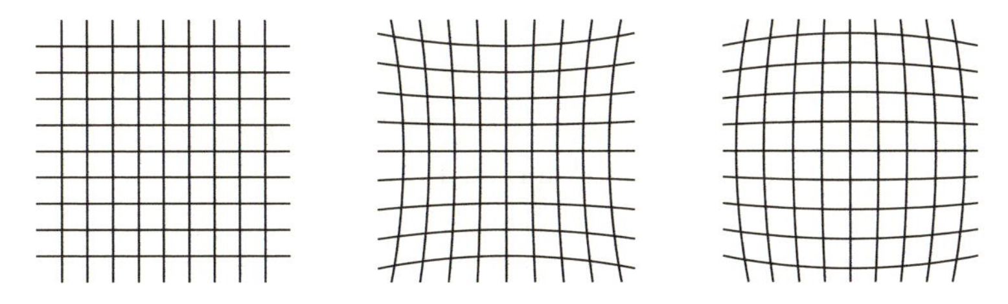

Figure 5.8: Distortion. An image of a regular grid of lines (left) should keep the lines straight. Instead the image will be altered by pincushion distortion (middle) or barrel distortion (right).

There are limitless different types of geometrical aberrations, most of which don't have names, and it is important to make sure the telescope is designed so that they are all but non-existent – something which is relatively easy with modern software. Once carefully designed, we then have to make sure the optics are made according to the shapes we want. In later chapters we will give a rundown on the basic techniques involved in testing and fabricating the mirrors. But even with a perfectly designed, shaped, perfectly smooth optics, we still have one more source of aberrations – misalignment. Just as with any other finely tuned instrument, it is important to make sure that all the parts are in the correct place for the best performance. As a simple example, imagine that the recording medium (which we'll assume is film for the sake of this discussion), is put in the wrong place – either too close to or too far from the secondary. This will cause the image to be blurred, because, just as with a camera, there is only one location for the film to obtain a focused image – any other position will give blurry images. Likewise, if we were to tilt one of the mirrors, or shift them sideways (called decentring) this would also introduce some blurring. The bottom line is that we must ensure that the mirrors and film all lie at the correct distances from each other and are not tilted or decentred. And to think that it all looked so simple in the diagrams in Chapter 2!

Field of view

Telescopes, like lenses in a camera, are designed with specific capabilities which are aimed at particular types of research. For example, photographers will use a telephoto lens in order to get a close-up of an athlete on a playing field. However, such a lens is no good for getting a wide-angle shot of the entire crowd in the stadium. In the same way, most telescopes are made for high magnification of distant scenes, but can only capture a small section of the sky at a time. In order to widen your field of view you either have to reduce your magnification or increase the size of your image plane. This alone would be difficult enough, but the problem is actually deeper than this: as we go to larger and larger fields of view, off-axis aberrations start to dominate. Designing a telescope with optics capable of minimising these aberrations is a real task. This is the reason behind the cost and complexity of professional camera lenses: they contain many component lenses of different shapes and sizes in order to maintain high image quality at the edges of the field of view as well as at the centre.

The problem is just as serious for reflector telescopes. Parabolic mirrors produce high-quality imaging over a narrow field of view, but they cannot be used for wide field imaging because off-axis aberrations (mostly coma) will appear. The solution to this problem was discovered by an Estonian-born optician, Bernhard Schmidt, in 1930. It was well known that coma is less of a problem for a spherical mirror, permitting imaging over a much wider field. But spherical mirrors are no good as they suffer from spherical aberration. Schmidt's solution was to introduce a thin piece of glass, specifically designed to compensate for the spherical aberration in front of the primary. The resultant telescope has a field of view ten times greater than that of other traditional telescopes. Furthermore, an added benefit is that we now have a spherical mirror which is much cheaper and easier to fabricate than a parabolic mirror.

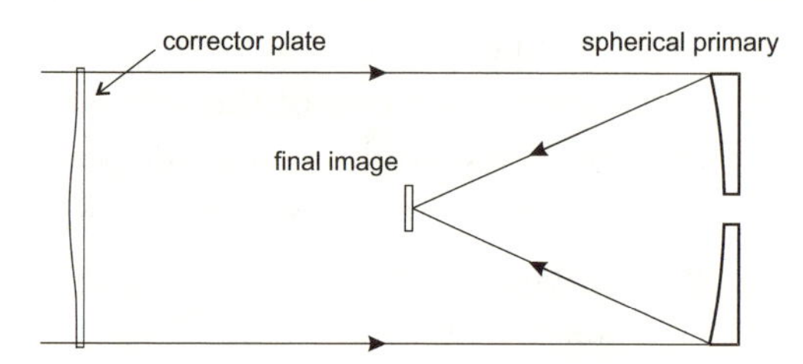

Figure 5.9: Classical Schmidt telescope. Light passes through a glass corrector plate located in front of a spherical primary mirror which then produces an image between them. The curve of the corrector plate is greatly over-exaggerated for clarity.

In its simplest form (the 'Classical Schmidt'), the image is formed in front of the primary mirror (see Figure 5.9). This is not very convenient for viewing by eye as it would require sticking your head in front of the primary. However, for astrophotography a camera can be placed at the location of the image. A more convenient design is the Schmidt-Cassegrain which has a convex secondary mirror designed to produce a final image behind the primary (Figure 5.10). This design has become the standard telescope for amateurs as the primary is cheap to make. The Schmidt corrector plate can be used as a support for the secondary and has the further practical benefit of providing a natural seal which protects the mirrors against the elements (dust, moisture, fingers, etc).

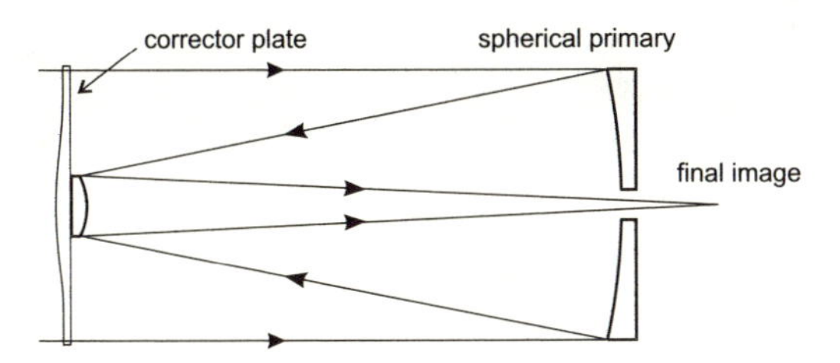

Figure 5.10: Schmidt-Cassegrain. Similar to the Classical Schmidt, the light is reflected off a secondary convex mirror to produce an image behind the primary.

For the professional astronomical community, the Schmidt telescope has many benefits which we will discuss in more detail later on. They are extremely valuable for all-sky surveys and catalogues as well as comparative studies for discovering temporal variations in stellar sources. It may seem odd, then, that the largest one is a mere 1.34 m in diameter. The reason is that beyond this diameter, the corrector plate becomes expensive to manufacture and cumbersome to mount. Recently, however, a clever design has emerged which uses a reflective Schmidt corrector instead of a refractive one. The Large Sky Area Multi-Object Fiber Spectroscopic Telescope (thankfully shortened to LAMOST) is being built by Chinese astronomers in the Yanshan mountains with a proposed completion date sometime in 2007. The use of mirrors and a completely open design gives it a 5-degree field of view with an incredible aperture of 4 m.

Air turbulence

So by now we are starting to come to grips with what it takes to make a perfect telescope. With an understanding of lens and mirror technology, along with a little time and money, we can make a fairly good instrument. But when we set it up in our backyards, we immediately notice that the images are blurry. In fact a star may end up looking something like that shown in Figure 5.11, taken by the 4.2-m William Herschel telescope. The star appears to be far from the Airy spot we were expecting and more like a fuzzy blob. Worse still, it moves about from one moment to the next. So what happened? The atmosphere, that's what.

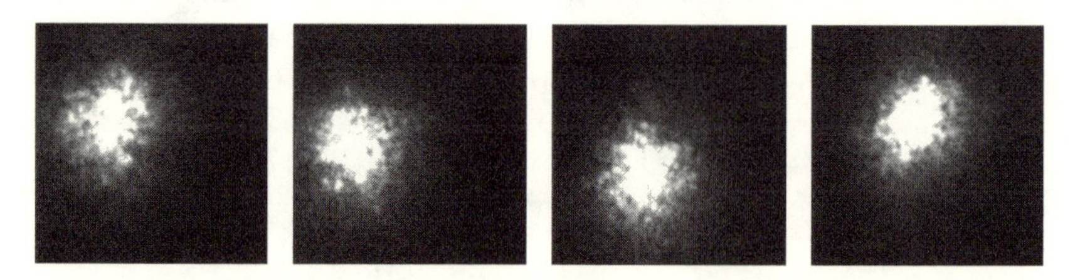

Figure 5.11: Four consecutive images of Betelgeuse taken 30 milliseconds apart. Notice how the image moves around from one moment to the next – in a timed exposure the final star image would thus cover a much larger size. Compare this with the diffraction-limited image taken by the 2.4-m Hubble Space Telescope in Plate 1. Courtesy: Blackett Laboratory, Imperial College.

The amount of energy absorbed by the atmosphere depends on many factors including latitude, time of day, cloud cover, humidity, temperature, altitude, pressure, aerosols/dust – as well as the albedo (reflectivity) and roughness of the ground terrain. The end result is that some parts of the atmosphere tend to be heated differently from others. Even a seemingly clear sky can have a lot of motion and differential heating taking place, completely invisible to the naked eye. Warmer air will refract light less than cooler air, and a plane wave of light passing through the atmosphere will bend more in some places than in others as it encounters different thermal pockets of air. By the time it reaches ground level, a wavefront can be severely distorted. This effect can be seen when looking through the layer of heated air just above ground level on a hot day, distorting distant objects into shimmering mirages. And at the risk of spoiling the romance somewhat, turbulence is also what causes stars to twinkle.

The extent of atmospheric distortion is often called 'seeing' by astronomers and can be defined as the largest aperture over which a wave is still diffraction-limited. To understand this, look at the aberrated wavefront shown in Figure 5.12. A large telescope will gather a wavefront with many portions exceeding the $\lambda/4$ limit, while a smaller telescope is less affected. We can now see that there will be some maximum diameter for the telescope primary such that the atmospheric distortion can stay below this limit. For even the best locations, this diameter turns out to be around 20 cm. A more aberrated wavefront from a more turbulent atmosphere would obviously limit the aperture even further.

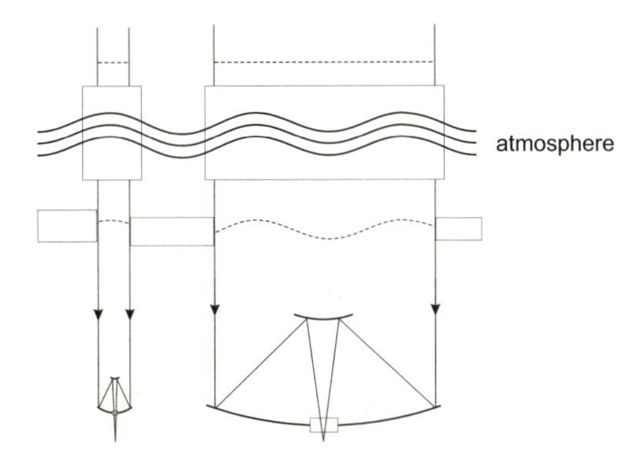

Figure 5.12: An initially flat wavefront (dashed) from a distant object in space will become aberrated passing through the atmosphere. A small telescope may not be affected by this, as the effect may be less than a quarter of a wave-length. A larger telescope will 'see' a wavefront aberration exceeding the diffraction limit.

Because of turbulence, then, it would seem that we cannot hope to get any improvement in resolution for a ground-based telescope by increasing the diameter of the primary beyond 15–20 cm. And this is indeed the case. But it would also seem that telescope engineers, knowing this, wouldn't waste their time building 10-m telescopes unless there was a reason. Right again. It all comes down to making the image brighter. The amount of light collected depends on the area of the primary or the square of its diameter. So, by doubling the diameter of a telescope, you collect four times more light and can see objects four times dimmer. If your resolution is limited by 'seeing', you won't improve the details, but you will be able to see objects which were previously too faint.

This may sound a little strange at first, but imagine walking into a dark room and looking around. You may be able to make out some objects, but generally there is not enough light to make out much. If you turn up the lights, you will gradually see more and more objects as they become brighter. Now, since the size of your eye has not changed, your visual resolution is roughly the same.[14] The reason you are seeing more is entirely due to the increase in light. The same goes for telescopes. Using larger and larger telescopes has the effect of 'turning up the lights' in the universe, so you will see fainter and fainter objects. This not only means more impressive images, but it also gives you more chance to analyse the light in more ways in order to give you more information about what the object is.

Chapter 6

Analysing the light

To this point we have considered how telescopes can be used to gather light and produce images of distant objects. Astronomers no longer actually look through telescopes, but instead attach all manner of instruments to them in order to take images and collect data about what they are looking at. Over the years, several techniques have been developed, each of which can be used to tell a particular tale.

Imaging devices – the camera

For two centuries, telescope images were viewed solely with the naked eye. We have already discussed the limitations of the human eye, and although the assistance of the telescope improves our observations of the cosmos, a huge leap in our discoveries came with the development of photography. The permanent nature of this recording medium made it possible to carefully analyse objects at a later date. Moreover, it also provided an indisputable record which removed many of the arguments associated with observer bias, imagination or fatigue. And because they often stay up all night, astronomers are particularly susceptible to the latter. Most importantly, however, the advent of photography finally made it possible to record objects much dimmer than those visible with the eye alone.

To get the maximum benefit from photography, a telescope must be moved to track an object throughout the night. Simply pointing a telescope in one fixed direction will result in the bright lines of star trails similar to those in Figure 1.2. The details of tracking methods will be given later, but suffice it to say that several possibilities exist which allow us to move a telescope to fix an object in the centre of our field of view throughout the night. With the addition of photography, astronomers could get images of very dim objects by building up a picture over time. Unlike the human eye, the chemical emulsion in photographic film does not have a refresh rate, so more light falling on the film over time will increase the image brightness. A galaxy viewed by the human eye through a telescope may appear as a dim blur, but using photographic film we can combine light over minutes or hours to form impressive images of objects too dim for the eye to see.

Photography involves creating a light-sensitive compound (such as silver halide crystals) embedded in a support material (gelatin) which is fixed on a substrate (either glass or celluloid film). When light strikes one of the tiny transparent crystals, it creates black silver atoms. After the proper exposure time, the light produces a weak latent image in the film. This image is then greatly enhanced with chemical processing in developer to produce a visible image. The negative is then made insensitive to further exposure by using a chemical called fixer. Telescope photographic negatives are generally black-and-white (or more correctly grayscale) rather than colour. The beautiful colour images we are generally used to seeing are produced by taking three such images through colour filters centred on the V, B and R photometric bands (see Appendix B for more details). An example of this colour recombination can be seen in Figure 6.1.

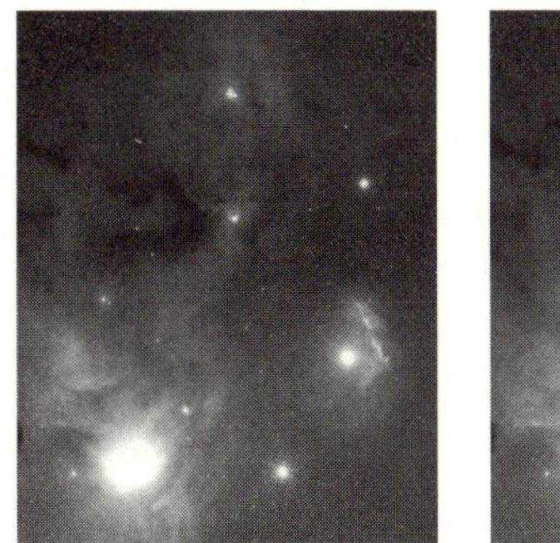

Figure 6.1: Three images of Rho Ophiuchi taken through red, green and blue filters combine to give the final image shown in Plate 2.

The first photograph of an object (the Sun) taken through a telescope was made in 1842. Initially such photographs produced limited improvements over visual records, but as the technology improved, so too did the resulting images. For over a century, photographic recording of telescope images followed this basic procedure. One improvement included gas hypersensitisation (or 'hypering'), which involves increasing the sensitivity of the silver halide crystals to light by presoaking the film in a chamber filled with a gas such as hydrogen (called 'forming gas'). Hypering greatly increases the brightness of images and permits the detection of dimmer objects. Further improvements included using films with smaller crystals (grains), which improved resolution. This might seem like a trivial accomplishment, but since smaller grains are less sensitive to light, it required some nifty chemistry to offset the sensitivity losses. By the mid-1980s, film technology was reaching its zenith, and the images produced by the world's largest telescopes were astounding.

Since the mid- to late 1980s, most telescopes moved on to electronic media such as charge-coupled devices (CCDs). Charge-coupled devices have an array of light-sensitive regions called pixels. A photon striking a pixel will cause a free electron to be produced. Multiple photons will produce multiple electrons and the charge in each pixel accumulates like drops of water in a bucket. When we wish to produce a final image, the amount of charge in each pixel of the first row of pixels is counted electronically. This charge is then discarded and the charge in the second row is transferred over to where the first row was (and third row to second, fourth to third, etc.). This is analogous to the transfer of water in a fireman's 'bucket brigade'. By this method we can eventually 'read-out' the charge in every pixel and produce an image.

Today, this type of digital recording has become the standard for all telescopes around the world. The benefits of this method over photographic film were slow to appear initially, but now far greater sensitivity can be obtained from such media and there are overwhelming gains in convenience. CCD chips do not require constant replacement (in a darkroom), nor chemical processing. This decreases downtime, allowing immediate access to results and permits the placement of the recording material deep within complex instrument packages. It is interesting to note that in the case of commercial cameras, film is still a superior recording medium since it has far greater resolution than digital media. However, the convenience issue outweighs this benefit for all but professional photographers.

CCDs are the standard form of digital recording medium, but variations exist which improve on sensitivity and speed, such as complementary metal-oxide semiconductor (CMOS) chips. The astronomical CCDs incorporated into the world's largest telescopes are far superior to the commercial-quality CCDs in a digital camera, with 10 to 100 times the number of pixels and much greater efficiency – approaching one electron produced for each incident photon. More importantly, they have much greater dynamic range (also known as 'well-depth'). Returning to the bucket analogy, we can think of this as the depth of the buckets, which ultimately permits the recording of much finer grayscale gradations. If you have a very deep well, you can have either a few drops of water (a dim object with few photons of light) or a large amount of water (a bright object with many photons) all recorded in the same image. A large dynamic range also makes it possible to observe minute differences and changes in object brightness.

Perhaps the most important property of astronomical CCDs is their extremely low readout noise. When we finally wish to detect the charge contained in each pixel, there is often some electronic noise which corrupts the process, producing slight imperfections in the calculated value. This 'readout noise' is much like trying to measure the amount of water in a bucket while it is raining. With astronomers particularly concerned with dim objects, this is an even larger problem, so their CCDs are specially manufactured in such a way as to keep 'noise' to a minimum.

Today's digital cameras have between 3 million and 10 million pixels, and are about the size of a postage stamp. The CCDs in advanced telescopes are significantly larger. They typically have around 250 million pixels and are about 30 cm square. As we will see later, future systems are being designed in the 1 gigapixel range. Unlike camera CCDs, these cannot be fabricated as a single chip as even the largest single chips have only 20 megapixels. Instead, the camera is built up from a mosaic of many smaller arrays. These are stacked as close together as possible in order to minimise the data lost in the small gaps between them. As you can imagine, with all these advanced features, the CCD in a typical large telescope is often custom made and as such is a very expensive item.

The images we see in publications, and those that astronomers actually use, rarely come straight from the telescope. For scientific research they have to be calibrated to take into account the idiosyncrasies of the telescope and in the CCD itself. For example, the individual pixels in a CCD chip may have different noise and sensitivity characteristics. One technique is called dark subtraction in which an image is taken with the shutter closed. This will record the random electronic noise which would have built up in our pixels even without light being present. This background is then subtracted from our final image. The second major calibration technique is called a flat-field frame. This is simply a record of a uniform background (for example, the twilight sky). Ideally in such an image, if all pixels were to 'see' 200 photons each, they would all record that number. In reality, one might record 195, while another records 199, etc. A flat field will allow us to correct for these differences in pixel sensitivity/ response. Once again, this information is used in adjusting the values of pixels in the final image. Figure 6.2 shows just how useful this technique can be.

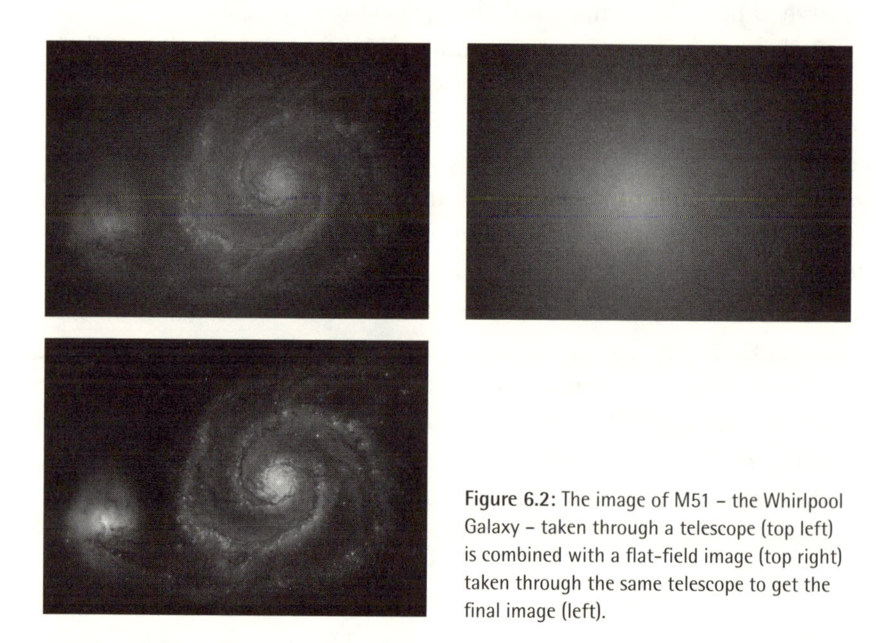

Figure 6.2: The image of M51 – the Whirlpool Galaxy – taken through a telescope (top left) is combined with a flat-field image (top right) taken through the same telescope to get the final image (left).

Astronomers use a variety of other methods for improving image clarity. This can involve 'stacking' multiple images to reduce noise – something that will be discussed in greater detail in Chapter 14. The types of techniques used depend heavily on the applications, and astronomers have to take care to ensure that they do not introduce spurious signals into the data. However, for press releases, images are generally manipulated to a greater extent. For example, streaks from cosmic rays, aeroplanes, meteors and satellites will be removed to avoid confusion for the non-technical viewer. If CCD mosaics have been used, there will often be dark lines from the small gaps between individual CCD chips, so these are removed as well. After this there may be further manipulation in the form of cropping, distorting, colour enhancement, contrast modification and even inverting to make a negative. In the case of infrared and ultraviolet telescopes, false colours are incorporated to highlight certain aspects to the viewer.

So does this mean that what we are seeing is the real deal or not? This turns out to be a surprisingly complex question. To see why, take a look at the image of Rho Ophiuchi shown in Plate 2. Remember that the human eye will not collect nearly as much light as a large telescope, so even if we were quite close to this nebula, the light levels might be so low that our image is built up from our retinal rods rather than our cones. Since rods essentially see in grayscale, this might mean that the nebula would appear much less colourful than shown here. In the case of Plate 2, the photographer (David Malin) has made it a large part of his life's work to ensure that the images we see are accurate representations of the objects as they might appear to us if we were to visit them. But given the subjective nature of vision, there is still much uncertainty.

Most of the data we see coming from telescopes is in the form of images. 'Happy snaps' of colourful planetary nebulae are great for calendars and press releases, but taken alone they don't actually provide a lot in the way of information about the physics of what is going on. Take a close look at the image in Figure 6.2. There is nothing to tell us how far away this galaxy is or how large it is. Nor can we tell what it is made up of or whether or not it is moving. I could tell you that this galaxy is made up of billions of stars in a disk several thousand light-years across, and that it lies some 31 million light-years away – well removed from our own galaxy. I could also say that it is mostly made up of hydrogen and helium, but there is nothing at all in this photograph to suggest this. I could tell you that stars on the edge of the galaxy are moving at about 500 km per second, but again you would be taking this on faith. By looking at the image, there is nothing to suggest that any of these numbers are correct.

So how do we know so much about these objects? The answer is that we have other techniques which we can bring to bear on the problem. Most telescopes are designed with a range of complex instruments that can be attached to better analyse the light in different ways. Here we will examine techniques such as spectroscopy, polarimetry, photometry and other concepts which can be used to impress people at parties.

Spectroscopy

One of the most useful astronomical techniques we have is spectroscopy. Spectroscopy involves taking the light from an object and spreading it out into its spectrum. An analysis of this spectrum can tell us about the motion of the object as well as about its material make-up and temperature. In order to understand how this works, let's return to the experiment conducted by Newton where he split 'white' sunlight into the full spectrum of colours using a prism. If we repeat this experiment, but spread the spectral light over a very large area, we may get the result as shown in Plate 3. For the most part, the entire spectrum is present, but we can see a few black lines indicating that some wavelengths are missing. Alternatively, we may notice that at some wavelengths the spectrum is brighter than at others. To understand what this means, we need to look at how light is emitted and absorbed by materials.

A simple model of the atom consists of electrons orbiting a central nucleus of protons and neutrons. Atoms of different elements contain different numbers of electrons, protons and neutrons. You could think of the electron as being much like a satellite orbiting the Earth. In this view, the electron has more energy the further away it is from the nucleus (just as it takes a more powerful rocket to get satellites further away from the Earth). In 1913 a Danish physicist, Niels Bohr, postulated that an electron is only 'allowed' in certain orbits or energy levels. The electron can only get from one energy level to another by absorbing or emitting a particular (quantised) amount of energy. For a given atom (e.g. hydrogen), the energy levels are fixed, so the quantised energy jumps between them are also fixed. Since energy is related to wavelength (see Appendix B for more details), the upshot is that jumps between these bound energy levels can only take place with the emission or absorption of well-defined wavelengths of light.

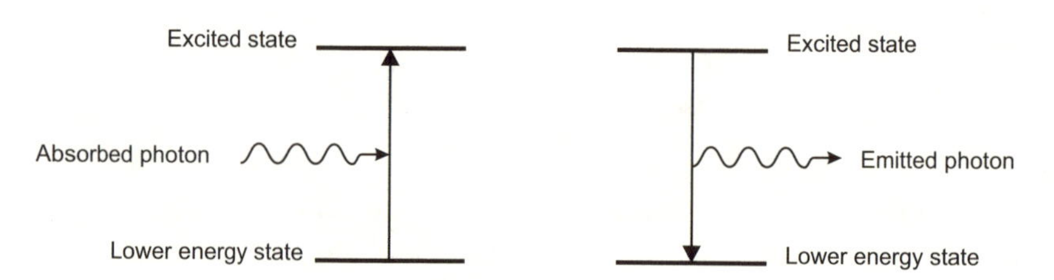

Figure 6.3: Examples of absorption (left) and emission (right) of photons taking an electron from one level to another. The energy levels are fixed, so the energy of the absorbed and emitted photons are also fixed.

So let's say we heat up a collection of hydrogen atoms in a glass tube. This will excite all the electrons to higher energy levels. At some later time, the electrons will relax back down to lower energy levels, by emitting light at certain fixed wavelengths (see Figure 6.4). Thus if we take a close look at the spectrum of light emitted by this glowing tube of hydrogen, we will see only certain colours of light present. Note that these 'emission lines' are not just in the visible regime, but can extend through other regions of the electromagnetic spectrum. As well as emission spectra, we can also obtain absorption spectra. If we illuminate a collection of hydrogen atoms with a white light source, we will notice that the light on the other side has certain wavelengths missing as the particular wavelengths are absorbed. These lines will coincide with the positions of the lines in the emission spectra for the same substance.

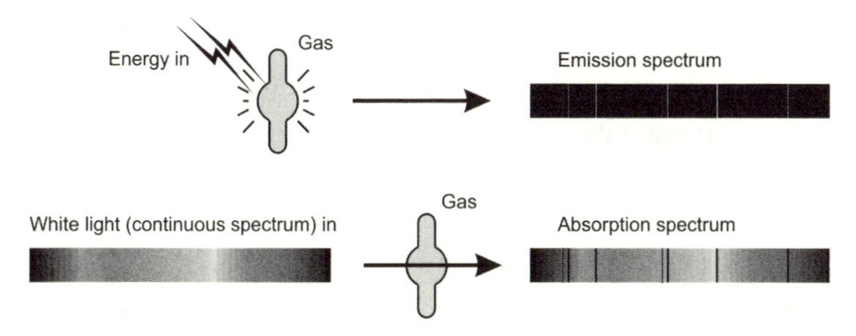

Figure 6.4: An emission spectrum (top) is produced by exciting atoms in a gas to emit certain wavelengths of light. An absorption spectrum (bottom) is produced when a complete white light spectrum is sent through a gas that absorbs certain wavelengths.

The key here is that different substances have different energy levels, and so produce different spectra. For example, in the case of sodium, there are virtually no bright emission lines except for a particular yellowy-orange line (actually two lines very close together). If you heat up sodium you can get a large amount of the emitted light at this wavelength and little at others. This efficient conversion of electricity into visible light is one reason why we use sodium in the distinctive orange street lights.

A spectrometer or spectrograph is an instrument designed to spread the light from an astronomical object into a measurable spectrum. The principle is fairly simple: light from the telescope is sent through a slit or an aperture to isolate the object in the field of view that we wish to analyse. This light is then directed onto something which splits the light up into its component wavelengths in a process called dispersion. While Newton used a prism for this purpose, prisms tend to have limited dispersion, so they can only produce a small spread in the spectrum. Astronomers prefer to use diffraction gratings to get a wide spread in wavelengths such as that shown in Plate 3. This in turn lets us identify very fine details in the spectrum.

A diffraction grating is an optical element in which there are many regularly spaced, microscopic lines or grooves. Reflecting light off this grating will disperse it into its distinctive spectrum. You have come across the dispersive effect of a diffraction grating in an everyday compact disc. In this case, the grooves are a series of pits in a regularly spaced spiral arrangement, but the result is the same. The diffraction gratings used in most telescope spectrographs consist of a series of perfectly parallel grooves in a reflective metal surface. Just as larger diameter primaries give better image resolution, so too do larger diffraction gratings give better spectral resolution. As a result, spectrograph gratings are often up to half a metre in diameter, have millions of grooves, and can separate out wavelengths which are fractions of a nanometre apart. In turn, better resolution allows us to separate out the absorption and emission lines from widely different substances which may be very close together.

Once again, the advent of digital recording technology has meant that the typical spectrum obtained from a spectrograph looks nothing like those presented here. Just as astronomers never look through their telescopes any more, they will also go their entire careers without ever seeing the colourful spectra presented here. Instead, the spectrum is projected onto a long, narrow CCD. From this the astronomer gets a readout in the form of a squiggly line on the computer screen like that shown in Figure 6.5. The wavelength at each location along the line is precisely calibrated, so we now have a record of the brightness of a spectral feature compared with wavelength. Spectra recorded by state-of-the-art spectrometers will have much more detail (higher resolution) and reveal much finer spectral features than the example given here, but the general appearance is the same.

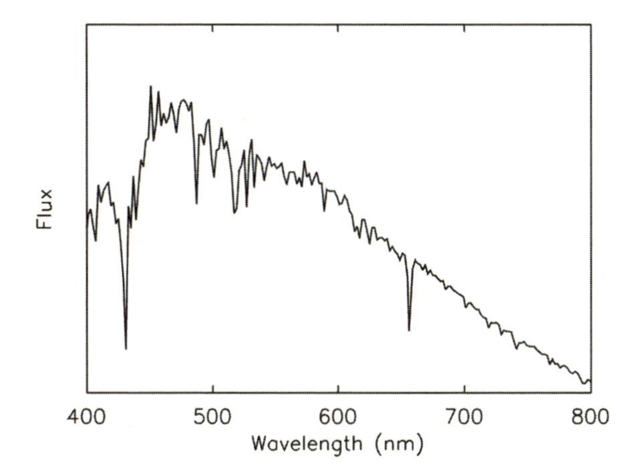

Figure 6.5: The spectrum as an astronomer sees it – in this case the characteristic emission spectrum of a typical star like our Sun. The dip at 656.3 nm is the characteristic 'H-alpha' red line from hydrogen.

At major telescope facilities, a good spectrograph is just as important as an imaging camera. Astronomers gather light from a distant object, and by examining the spectral lines, they can deduce the presence (or absence) of certain substances. Absorption lines indicating the presence of hydrogen were first identified in the spectrum of the Sun by Fraunhofer in 1814. Plate 3 shows an image of many absorption lines in the solar spectrum which can be ascribed to several different elements. In fact, this is how helium was first discovered. In 1868 a French astronomer, Pierre Janssen, was observing the spectrum of a solar eclipse when he noticed a yellow line which was not associated with any other known element. He theorised that this was produced by a new substance which he called helium after Helios, the Greek god of the Sun. The element was not isolated on Earth until 1895. This is significant as it illustrates how the observation of a distant astronomical source can lead to discoveries relevant to our life here on Earth.

Spectroscopy can also be used to measure motion and temperature. For motion we use a property called Doppler shift. When there is emission from an object which is moving along our line of sight, this wavelength will be shorter or longer than for a stationary object. For objects moving towards us, the wavelength is shortened towards the blue end of the spectrum, and for receding objects, the light will be red shifted. You have noticed this effect with sound waves, but you probably no longer hear it. Try this: pretend you are imitating the sound of a very fast car passing by. Listen carefully, and you will hear that you make a high-pitched sound (shorter wavelength) as it approaches, changing to a low-pitched sound (longer wavelength) as it passes. With light the same thing happens, but the Doppler shift only becomes large for objects moving close to the speed of light. Still, with a sensitive spectrograph we can measure wavelength very accurately, so we can 'see' when an object is moving even quite slowly. For example, we might examine the spectrum of light from a star, and find that the distinctive emission lines from hydrogen are shifted slightly from the wavelength they are supposed to be. By measuring the shift, we can measure the speed at which the star is approaching or receding. We will see later on how these velocity measurements have been critical in all manner of astronomical discoveries.

Let's now look at measuring temperature. Until now, we have considered the spectra from atoms and molecules. If we heat a substance enough, an electron may be able to leave the substance completely. This leaves us with an ion – a charged atom or molecule with one or more electrons missing. An ion has a different emission spectrum than an atom, so once again we can detect the presence of ions by their unique spectral lines. Spectroscopic measurements of this type had long shown that the Sun had a 'surface' temperature of around 5500 degrees Celsius. However, several emission lines have been discovered in spectroscopic observations of the corona (the outer 'atmosphere' of the Sun) which indicate the existence of Fe 13+; an iron atom which has had 13 of its electrons removed. Now, it takes a large amount of energy to remove a single electron from an atom. To remove thirteen electrons from an atom such as iron requires huge amounts of energy. In fact, calculations show that the corona must have a temperature of a few million degrees! Just try to imagine the surprise (and disbelief) when these measurements were first published. Common sense would suggest that the temperature would decrease as you moved further away from the surface of the Sun, but it would seem that this is not the case. Something is causing a dramatic increase in the temperature at the outer reaches of the Sun. The cause of this heating is still not entirely clear, but is most probably the powerful solar magnetic fields.

Photometry

Photometry is the process of measuring the amount of light coming from an object. This can be done simply by looking at the light levels in individual pixels of a CCD, or by using specially designed light sensors. Photometry has many uses. When a supernova explodes in a distant galaxy, a measure of the change in light output can tell us what type of star exploded. We can also use this technique to observe variable stars – stellar objects which change in luminosity over days or months. These stars are useful in determining distances to other galaxies – a subject which will be discussed in more detail in Chapter 9. Photometry can also be used to measure 'light curves' of asteroids, which can tell us how fast they are rotating, even if we can't resolve the object. To see how this works, imagine a white car tumbling in space a long distance away. Generally sunlight will reflect off the white surface to give a large signal. When the car turns so that the dark undercarriage faces us, the light signal will drop to a lower value. By observing the way the signal changes from high to low, we can measure how fast the car is spinning. The same process works for measuring the rotation of asteroids.

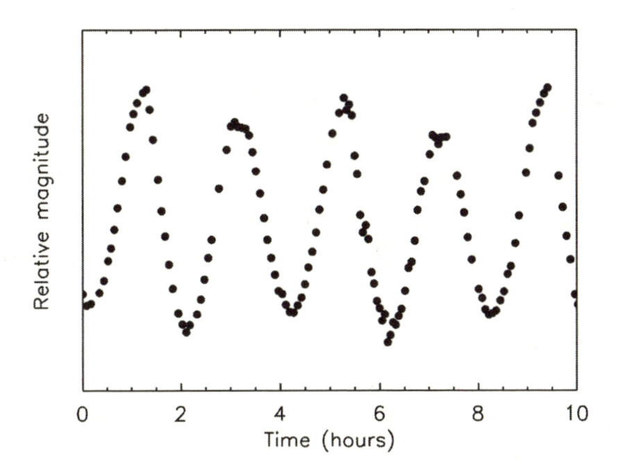

Figure 6.6: The above plot shows a series of measurements of the brightness of the asteroid 347 Pariana. This graph is a combination of two separate observations over two separate nights, and shows the rotation rate to be just over four hours long. Courtesy: Lt. Col. Wetterer, USAF Academy.

Recently a Canadian consortium launched a 15-cm diameter telescope called MOST (Microvariability and Oscillations of Stars) into low-Earth polar orbit. This suitcase-sized satellite is capable of monitoring brightness of light from a target to a precision of around a micromagnitude or around one part in a million. In a recent talk, a mission scientist Jaymie Matthews put it this way: if you were to see the Empire State Building with every light in every office turned on,

MOST would be able to detect a change in the brightness you'd get from drawing the blinds on one window by a mere three centimetres. With this sort of sensitivity you can observe a multitude of cosmic phenomena. For example, by monitoring the light from a star the photometer can 'see' changes in brightness caused by the presence of sound waves ringing throughout the stellar sphere (stellar seismology) as well as sunspots moving over the face of these stars. And all this can be done without even being able to resolve the disk itself. The future is bright (literally) for this instrument as it turns its gaze towards other stars, finding planets as they dim the light of the star by passing in front. MOST can go even further by detecting the reflected light from the planet just before it passes behind the star, which can give us an idea about what sort of atmospheres they may have.

Polarimetry

Light is a wave which has a periodic variation in electric (and magnetic) fields. Generally when light is emitted from an object, it comes out with the electric field oscillating in all possible directions. However, in some cases we can restrict this oscillation to one particular direction only. This is polarised light. A typical method of doing this is by passing the light through Polaroid film. This restricts light much like a wave in a hose passing through a picket fence. You can get the hose (i.e. the wave) to oscillate up and down, but the fence stops oscillations from side to side. When two polarisers are arranged at right angles to each other, all light is blocked out.

Polarised light can be produced by many natural processes. For example, light reflected off non-metallic surfaces will be polarised. For this reason, Polaroid sunglasses are ideal for blocking out glare from sunlight reflected off car windows and water. When light is scattered off molecules or dust at certain angles, it can also become highly polarised. You can see this for yourself if you look at the blue sky through Polaroid sunglasses. At an angle of ninety degrees to the Sun, the polarisation is maximised. If you look at a patch of this sky and rotate the sunglasses, you will see the sky go from dark to light as the film is alternately aligned with or against the direction of polarisation. Astronomers take advantage of this in order to see if light from an object is polarised or not. For example, if we look at the nebula in Plate 2, we may not know if the luminous gas and dust clouds are being lit by some outside object (a nearby star, perhaps) or if the light is being generated internally. These differences can often be spotted in the nature of the polarisation of the light, which in turn can help astronomers determine the origin and make-up of these clouds.

From the short summary given in this chapter we can see that astronomical telescopes are more than just big devices to produce pretty pictures. A lot of effort and expense goes into the design and construction of the instruments, and the groups who build them often come up with some clever acronyms. Names such as ELMER, FLAMES, VIRUS, PEPSI, HIPO, LUCIFER, KAOS, PEPPER, OzPoz and GIRAFFE demonstrate that they also have vivid imaginations. Now that we have an idea of how the light is ultimately analysed, we will look at a different way in which the light can be collected in the first place.

Chapter 7

Interferometry

Until now, we have considered telescopes consisting of a single primary. In this chapter we will examine the possibility of using multiple apertures to collect light and then combining it to form a single image. We will see that this makes it possible for many smaller telescopes to give a similar performance to a much larger, single telescope. At this point I should mention that this chapter involves some pretty advanced concepts, which may be heavy reading. While interferometry is the next big thing for telescopes, it is a complex subject, so some readers may want to skip over this material. This can be done without any fear of misunderstanding discussions in later chapters.

Interference – how light waves combine

In the historical summary given in Chapter 2, we saw that Newton discovered many properties of light in his treatise *Opticks.* While Newton made numerous advances in this field, he did make his share of mistakes. For example, it was Newton's firm belief that light consisted of particles – the so-called corpuscular view. For over a hundred years this view prevailed, mostly because of the difficulty scientists had in opposing the views of such a formidable physicist. However, evidence was being gathered at a remarkable rate which did not sit well with the idea of light as a particle. For instance, refraction is simple to explain if we treat light as a wave, but much more difficult to understand as a particle phenomenon. One of the most critical clues for light waves was gathered by Thomas Young.

Thomas Young was born in Somerset, England in 1773. An example of a true child prodigy, he was (according to his own account) able to read at the age of two, and had read the Bible more than once by age four. He spoke eight languages, studied a variety of scientific disciplines, and was a key player in the decoding of the Rosetta Stone. Young was the first to realise that the eye focused by using muscles surrounding the lens to change the curvature and hence the focal length of the lens. He studied many optical phenomena throughout his life and increasingly began to believe that light was a wave. Furthermore, he began to suspect that different colours were in fact different wavelengths of light, just as notes of different pitch are different wavelengths of sound waves. Ironically, much of his reasoning came from an analysis of the Newton's rings phenomenon described in *Opticks*. In 1801, Young devised an experiment which involved combining light with a common source, from two different locations. He did this using two slits (or holes) lit by a single light source, as shown in Figure 4.1. On the screen located some distance away he noticed the appearance of light and dark bands (fringes). To explain this, Young had to rely on a wave theory of light.

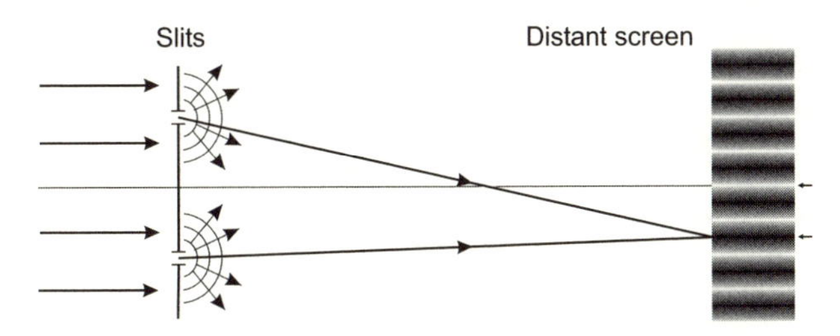

Figure 7.1: Young's Double Slit Experiment. Light spreads out through two closely spaced slits. On a distant screen, light from each slit combines to give a pattern of light and dark fringes.

When one light wave meets another, they combine in a unique way. Normally we are accustomed to the idea of one plus one equals two, but in the case of waves, we can get one plus one resulting in any value between zero and two. Let's consider two simple waves as shown in Figure 7.2. If two waves combine 'in phase' in such a way that a crest may coincide with a crest and a trough with a trough, they will add together. This is called constructive interference and will produce a doubling in the amplitude of the wave. Alternatively they may combine 'out of phase' so that the trough of one wave coincides with the crest of the other. This is known as destructive interference and results in the waves cancelling each other out. Constructive interference gives a brightening of light and destructive interference results in no light (darkness). For combinations of waves between these extremes, we will get something between bright and dark.

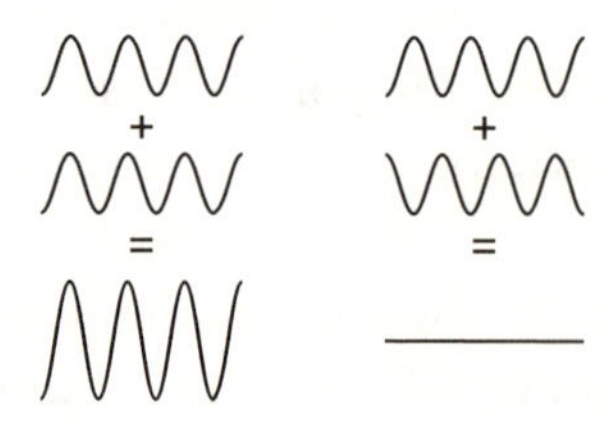

Figure 7.2: Wave interference. On the left, the top two waves are in phase, so combine to give a wave of double the amplitude. On the right, the two waves are out of phase and thus cancel each other out, giving no wave and hence no light.

Now let's look at the double slit experiment once more. The single source of illumination ensures that light leaving each slit will always be at the same point in the wave cycle – i.e. they start out in phase. At point X on the screen, midway between the two slits, light waves from each slit will have travelled an equal distance. This means that they will add up constructively to give a bright fringe. At some point Y, slightly off to one side, light from the upper slit will have travelled further than that from the lower slit. If it so happens that this difference in distance is exactly half a wavelength of light, the two waves will destructively interfere to give a dark fringe. As we move along the screen we will see that the difference in path length travelled by waves from each slit will alternately give light and dark fringes – just as Young observed (Figure 7.3). The importance of these fringes is that they proved that light was a wave. Using a corpuscular view of light, the combination of particles from two slits would just give an overall blur, not an oscillating pattern of fringes.

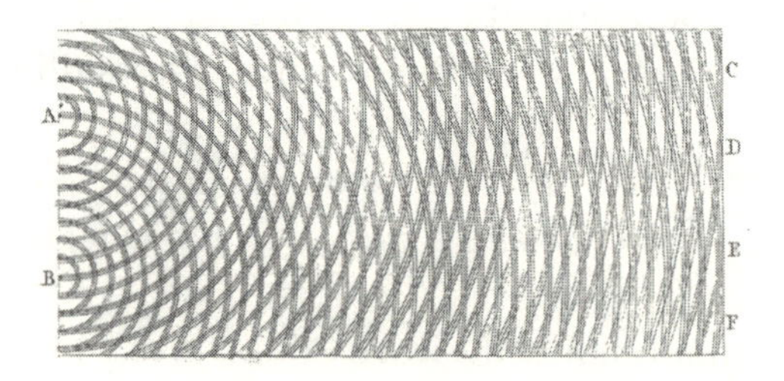

Figure 7.3: Young's explanation of wave interference in his double slit experiment. Circular waves emanate from two points A and B on the left and combine to give dark fringes at points C, D, E and F.

As mentioned in Chapter 2, Huygens had already described a wave theory for light in 1678, but with Newton subscribing to the corpuscular view, this theory was largely disregarded. By the end of the nineteenth century however, the phenomena of interference, polarisation, diffraction and refraction all had simple wavelike descriptions, but no acceptable particle interpretations. So by the time James Clerk Maxwell was born in Edinburgh, Scotland in 1831, the general consensus had swung firmly behind light as a wave. In his 1864 work entitled *Treatise on Electricity and Magnetism*, Maxwell presented four equations which succinctly described both electric and magnetic fields as being related in a single phenomenon called electromagnetism. A varying electric field will produce a magnetic field and vice versa. Furthermore, when a charge is moved, it will emit waves of mutually perpendicular oscillating electric and magnetic fields. From his theory, Maxwell calculated the velocity of the electromagnetic radiation as an expression dependent on constants of electric and magnetic field strengths. The resulting number though, is the speed of light, which led Maxwell to conclude that the electromagnetic radiation in question was in fact light (see Appendix B for more details). So now all was well in the world, and it seemed that there were no more mysteries to be solved in the matter. This couldn't have been further from the truth.

Michelson interferometer

Albert Michelson was born in Strelno, Prussia (now Poland) in 1852 and moved to the United States at the age of two. He grew up in California and Nevada and on graduating from high school decided he wanted to enter the US Naval Academy (Annapolis). Unfortunately, the two positions available for students from Nevada were taken, so at the age of 17 he travelled on his own to Washington. Upon arriving at the White House, he personally requested a presidential appointment – apparently, those were the days when you could just knock on the door and get such an audience – and successfully petitioned for a special dispensation. He flourished in the environment of the Academy, performing well in both mathematics and sciences in addition to being a champion boxer. However, he did have some problems with his military studies and was once admonished by the Superintendent, who commented:

> If you'd give less attention to those scientific things and more to your naval gunnery, there might come a time when you would know enough to be of some use to your country.

Fortunately for the scientific community it appears that he did not take this advice to heart. After graduating in 1873, he served at sea for two years before returning to the Academy as an instructor in physics. Unlikely as it may seem for someone who was to later win a Nobel Prize in physics, he never actually earned a formal degree, as they were not awarded at Annapolis until several years later. As he matured in the field, he began to perform some of the experiments which would make him famous – including an improved measurement of the speed of light. He resigned his commission in 1882 and from then on spent the rest of his life in academia.

In 1880, while on leave from the Navy at the University of Berlin, he developed a device now called a Michelson interferometer. A Michelson interferometer takes a beam of light and splits it up into two beams which travel at right angles to each other (see Figure 7.4). In his day, Michelson would use light of a single colour from a sodium or mercury lamp, but today we would use a laser (which is much easier – trust me). The beams are then reflected back along the direction they came from and recombined by the beamsplitter. The result is that we end up with two beams of light which have taken two different paths, but are made to overlap. As in the case of Young's double slit, and in fact any time two beams of light are made to overlap, we get an interference pattern of light and dark fringes. The significance of the fringes is that they actually tell us about differences in path length between the beams of light taking each path. If we move one of the mirrors even a billionth of a metre, the fringes will move dramatically. This makes it the ideal device for measuring extremely small distances, and this is precisely how Michelson used it. Together with Edward Morley, he used this device to demonstrate that the æther does not exist.

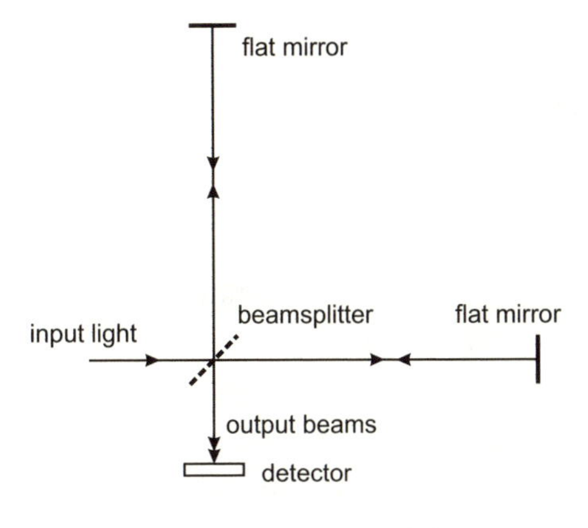

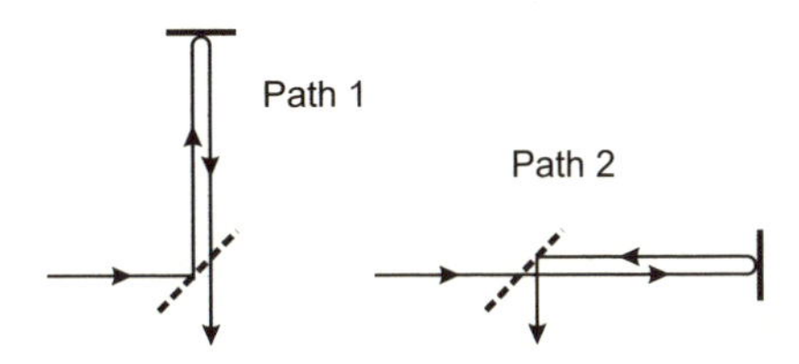

Figure 7.4: Michelson's Interferometer. A single beam of light is split into two, reflected back along the outgoing paths and then recombined. The bottom two images show the two different paths the beams can take.

The existence of the æther was proposed to explain the nagging problem of the transmission of light waves in outer space. The wave nature of light required that there be a medium through which the wave could propagate. Just as sound waves cannot travel in a vacuum (as they rely on the motion of atoms), it appeared that light too required some medium in order to get from one place to another. Physicists were quick to see a problem developing. Space was thought to be a vacuum – so how could the light from stars reach us? The natural explanation was that space must be filled with some all-pervading medium, called the æther. The existence of the æther was a real thorn in the side of physicists, as astronomical observations put extreme limits on its properties. For example, it must be very tenuous, or else the Earth would experience noticeable drag in its orbit, and eventually spiral into the Sun.

Michelson realised that his interferometer was a perfect device for measuring the properties of the æther. The Earth moves in an east-west motion around the Sun through this æther. Thus there should be a difference in the speed of light in this direction compared with some perpendicular direction (north-south, say). A difference in speed is the same as a difference in path length to the interferometer. So first the interferometer is aligned so that Path 1 is aligned north-south, while Path 2 is aligned east-west. Fringes are observed as the difference between the two paths. By gently rotating the interferometer by 90 degrees, Michelson expected the æther to lengthen Path 2, while Path 1 shortened by the same amount. This would show up as a shift in the fringes. It is important to realise just how sensitive this measurement is – the slightest vibrations will disturb the two arms so that the fringes disappear. Michelson was forced to perform these experiments in a basement at the Astrophysical Observatory in Potsdam. Even then he had to take measurements after midnight because people walking by up to 100 m away would affect the results. After many careful trials, Michelson was finally convinced that he was not seeing the expected fringe shift, and eventually came to the conclusion that there simply was no æther.

To the physics community, this discovery presented something of a good news/bad news predicament. The good news was that the concept of the æther could be tossed out – it had been a concept which no one really liked anyway. The bad news was that a new theory would have to be developed to explain how light propagates in a vacuum. This problem was not completely resolved until 1905 when Einstein showed that light can take the form of particles, known as photons. With this discovery, light became something of an oddity – it could be either a wave or a particle depending on how you observed it. This particle-wave duality ushered in the dawn of quantum mechanics and has since been shown to be a property of all waves and matter.

Michelson became the first American Nobel prize winner in physics (1907) for 'his optical precision instruments and the spectroscopic and metrological investigations carried out with their aid'. This seems to be a rather convoluted way of saying, 'Well, he did so much important work that we had to give him a Nobel for being so clever.' Most physicists at the time were in awe of the precision of Michelson's measurements and devices – they were often over ten times better than anything that had previously existed. For example, when he turned his attention to measuring the speed of light, the value had been determined by Foucault to be 298,000±500 km/sec. Throughout various experiments conducted over 50 years, Michelson refined this value to 299,796±4 km/sec – a 100-fold improvement in precision. This work ultimately led to the decision of the scientific community to redefine the metre in terms of the speed of light.

Michelson developed many other types of interferometers which are used for a variety of applications. As they are capable of measuring extremely small variations in distance, interferometers are ideal for detecting the smallest imperfections in optical components. If we were to have perfectly flat mirrors in each arm of a Michelson interferometer, we would expect to see nice straight fringes in our final pattern. If one of the flat mirrors is replaced by a mirror with some surface irregularities, we would see these as a bending in the fringes. In Figure 7.5 we can see the pattern from a poor-quality optic. Notice how the fringes can be used to produce a contour map of the lumps and bumps on the surface – in the same way as topographical maps show us the location and size of valleys and hills. Remember that the distance between the contours in this case indicates a change in 'height' of just $\lambda/2$, so the variations in height are extremely small. In fact, with some clever improvements, we can actually make this measurement billions of times more sensitive, and use such an interferometer to measure variations in the shape of space-time itself. This will be described in a later chapter.

Figure 7.5: Measuring surfaces. On the left is shown the interferometric measurement of the surface of a high-quality flat mirror. On the right is a typical bathroom mirror revealing a surprisingly distorted surface. Incidentally, the two small circular ring patterns on the left are due to dust – horrible stuff.

Michelson stellar interferometer

In 1890, Michelson devised a way to combine the light from a distant object, gathered by two separate apertures to produce an interference pattern. This device, known as a 'Michelson Stellar Interferometer' (or simply a stellar interferometer) should not be confused with the Michelson interferometer described above. It works like this: starlight is collected by two widely separated mirrors and directed into a telescope which focuses the light and combines the two beams at the focal plane (see Figure 7.6). You should think of this in the same way as the two slits in Young's double slit experiment (Figure 7.1). As in that case, we have a single light source (the star), sampled by two apertures (the two widely separated mirrors), and with the light from each overlapped, it gives rise to interference fringes. The distance between the mirrors (*D*) is called the baseline.

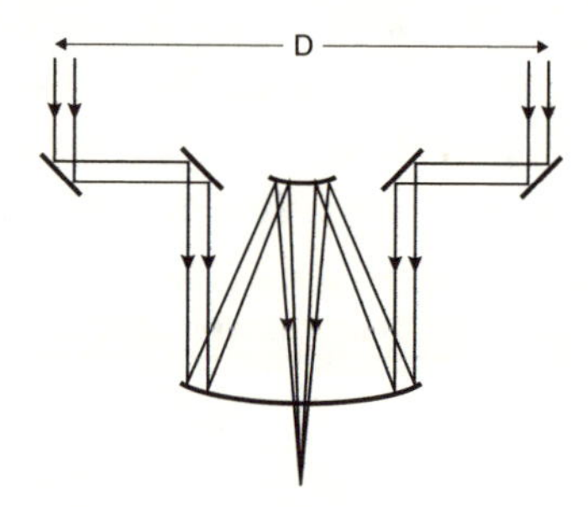

Figure 7.6: The stellar interferometer. Starlight gathered from two small, widely separated mirrors is combined in a telescope to give interference fringes. The resolution diameter (*D*) in this case is the separation of the mirrors, not the diameter of the telescope primary.

Michelson realised that the stellar interferometer had two benefits over a conventional telescope. Firstly, two small mirrors on a long boom are a lot easier to make and align than a huge telescope. Furthermore, smaller mirrors are less susceptible to atmospheric turbulence. Using a ground-based telescope with a single large primary, it would be all but impossible to resolve the disk of a star as a result of limitations imposed by atmospheric turbulence. With the stellar interferometer, however, the light is collected by small diameter mirrors, so the effect of turbulence is minimal. The main difference between a regular telescope and the stellar interferometer is that the 'image' of a star looks something like the one shown in Figure 7.7. It is the fringes that give us information about the size of the star itself. For a large separation of the outer mirrors, the fringes will have a high visibility. That is to say, they go from white to black. As we reduce the distance between the mirrors, however, the visibility of these fringes decreases so that they look more like light grey and dark grey. Decrease the baseline even further and eventually the fringes disappear altogether. At this point, we measure the mirror separation, and use the resolution formula to calculate the angular diameter of the star being observed.

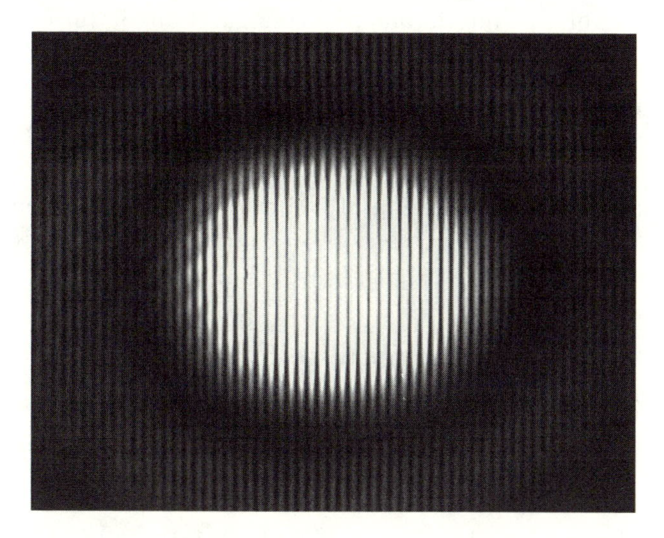

Figure 7.7: The 'image' of a star created by a stellar interferometer. The fringe separation is determined by the baseline, whereas the overall large Airy pattern is a result of the size of the small round mirrors.

In 1920, Michelson conducted an experimental evaluation of the idea on the 2.5-m Hooker reflector telescope at Mt Wilson Observatory. Michelson used a boom set on the top of the telescope to extend flat mirrors out to a baseline of 15 m. In the first successful test, the fringes observed from Betelgeuse were seen to disappear at a mirror separation of just 3 m. Taking a wavelength of 575 nm for the light, the angular diameter of the star was calculated to be $1.22\lambda/D = 0.23$ microradians. Multiplying this angle by the current value for the distance to Betelgeuse calculated from other observations (425 light-years) gives a diameter of almost 940 million km. This is just 6.2 astronomical units or 6.2 AU (see Appendix A for more information on this unit of distance). With this experiment, Michelson made the first ever measurement of the angular diameter of a star, and the method has since been applied to hundreds of other stars. It was not until 1996 that the first image was taken that resolved the disk of a star (something which Michelson himself thought would be impossible). Coincidentally the star was Betelgeuse, and the image snapped by the Hubble Space Telescope is shown in Plate 1. To get an idea of the scale of this star, its diameter is twice the size of the orbit of Mars. Betelgeuse is a red giant in the late stages of its life when (like many of us in our old age) it has become bloated and dim.

Imaging interferometry

As a result of the interferometric configuration, it turns out that the angular separation between fringes is $0.5\lambda/D$, where D is now the baseline distance. This means that the image resolution is automatically 2.44 times better than that given by a single primary mirror of comparable diameter. An adequate explanation of why this is the case would almost certainly warrant a reclassification of this book from 'popular science' to 'physics text' and incur the wrath of the publishers, so you will have to trust me here. It may seem, then, that this represents something of a free lunch – throw away most of the telescope mirror, and you can actually improve the resolution! As we all know, there is no such thing as a free lunch, and this case is no different. If we examine the set-up carefully we find that it only applies along the line joining the two mirrors.

Take a look at Figure 7.8 and you'll see what I mean. For a filled aperture, the telescope has a resolution of $1.22\lambda/D$ in all directions. For two small mirrors, the resolution is $0.5\lambda/D$ in the horizontal direction, but in the vertical direction the effective diameter is a much smaller d which is only as large as a single one of the small mirrors – i.e. not much at all. The bottom arrangement would produce the pattern shown in Figure 7.7. There is another problem. The two separated apertures do not collect all the image information compared with that of a filled aperture. As a result, images are not a true reflection of how they actually appear and some detail is lost. However, it is a useful technique for getting some gains in resolution without requiring the construction of huge telescopes.

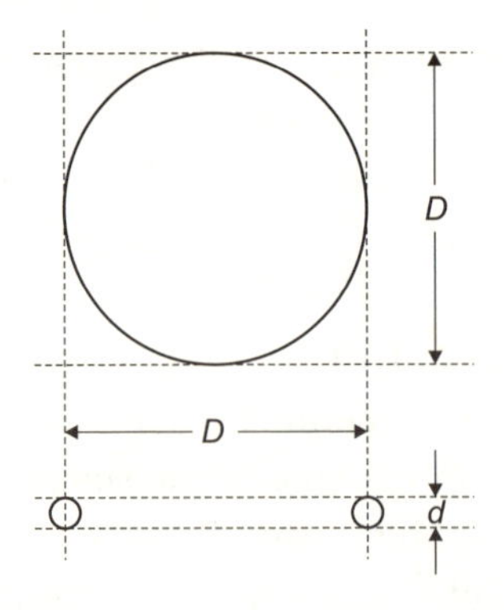

Figure 7.8: Resolution of the stellar interferometer. The filled telescope (top) has a resolution diameter D in all directions. The two-mirror interferometer (bottom) has an effective diameter D in the horizontal axis only, with a smaller aperture d in the vertical.

In the example given above, the light-collecting power has dropped dramatically as we now have very small mirrors. This limits the application of stellar interferometry to brighter objects, unless large mirrors are used. A second limitation comes from the problem of setting up the mirrors in the first place. While a boom attached to a telescope worked for Michelson, it was cumbersome and limited by the size of the dome. Added to this there was still the problem of loss of light-gathering power. A solution is to build two large fixed telescopes to give a much larger baseline. The light from the two telescopes can then be combined to produce the high-resolution fringes. An example of this approach are the twin Keck telescopes on Mauna Kea, Hawaii, shown in Figure 7.9 and Figure 7.10.

Figure 7.9: The Keck I and II telescopes can combine their light interferometrically to make high-resolution measurements.

The combination of light in this case is no longer the trivial matter of simply directing it into a single small telescope as it was with the Michelson stellar interferometer. The reason is that we no longer have a fixed boom, so as we track an object across the sky with separate telescopes, there will be a change in the path lengths of the light into each telescope. There must be a way of compensating for this difference in path length or else the high-resolution information will be lost. In the case of the Keck telescopes, this compensation is done by channelling the light from each telescope into tunnels beneath the building where mirrors on high-precision rails (called delay lines) can make one beam travel a longer or shorter path relative to the other before the beams are allowed to combine. This sounds pretty straightforward, but the paths must be matched to within a fraction of the wavelength of light, while the overall path difference changes by tens of metres over the course of a night's observations. In order to achieve this degree of control, the tunnels must be designed and built to remove the effects of air turbulence, and complex 'fringe-tracker' systems must constantly monitor the positions of the movable mirrors.

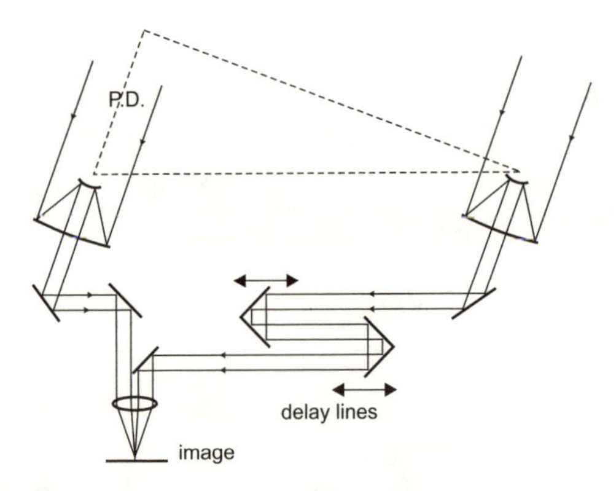

Figure 7.10: A schematic of the Keck Interferometer. The path difference (P.D.) between the light entering each telescope must be compensated for by changing the lengths of the delay lines.

Large telescope stellar interferometry has been the dream of astronomers for decades, but has only been made possible in the last few years with advances in optics and engineering capabilities. The Keck interferometer consists of the two telescopes, each with a diameter of 10 m, set 85 m apart, and is designed (at least initially) to operate in the infrared where the longer wavelength greatly eases the alignment and path-matching issues. Meanwhile, the Very Large Telescope (VLT) in Chile operated by the European Southern Observatory has four individual telescopes, each with a diameter of 8.2 m, which can be combined in pairs or all together. The first pair were combined for a baseline of 102 m in a 'first fringe' observation of Sirius on 17 March 2001. Future designs, however, aim to extend this technique to add more smaller telescopes dotted around the main telescopes. These so-called 'outrigger' telescopes will help fill in some of the blanks in the image and improve the overall resolution. As always, though, delay lines are required for each new telescope in order to ensure that all of the light is combined in the correct way.

Still, if you are patient, you could use such an interferometer to build up an image. Such telescopes are called imaging interferometers, or sparse array telescopes. With a telescope consisting of two widely separated apertures we can take a high-resolution 'slice' through the object. If we were to rotate the whole arrangement, we could take a series of image slices at this higher resolution through the object of interest. Given enough slices, this process could be used to build up a high-resolution image of the object. In this case, the individual telescopes are fixed, so the orientation of the baseline cannot be modified. Instead astronomers rely on the rotation of the night sky to achieve the same effect. Over time this technique can be used to create images with much better clarity than would be possible with a single large telescope.

Nulling interferometry

To this point, we have talked about combining light between two beams to produce fringes. The fringes are produced by having some angular separation between the two beams from the two smaller apertures. To see this, look at the final two beams as they are combined by the final mirror in Figure 7.6. There is some angle between the two beams which determines their size on the image plane; the smaller the angle, the wider the fringes. This is shown in Figure 7.11, where the angle is narrowed greatly.

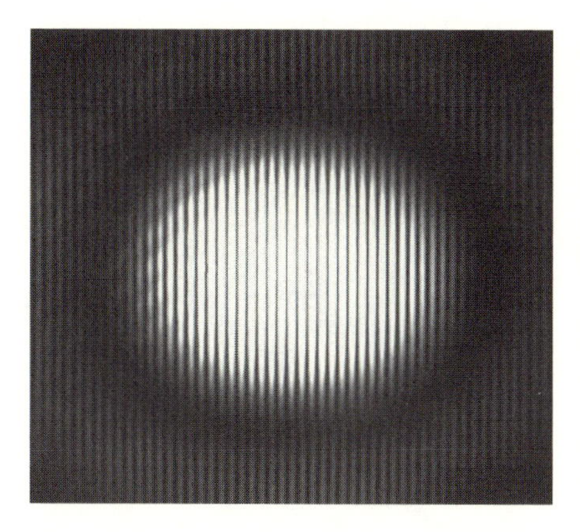

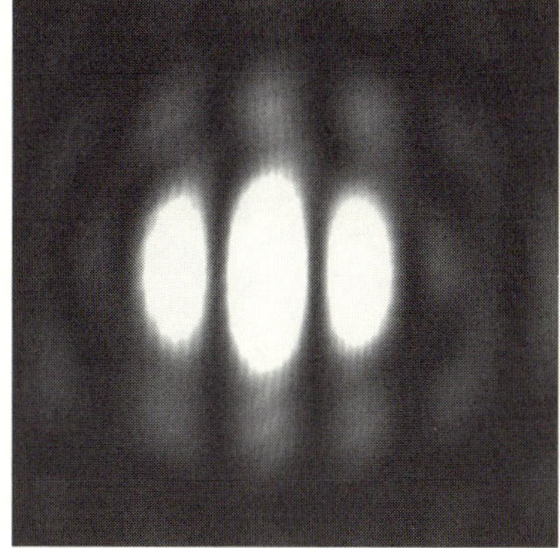

Figure 7.11: A smaller angle between the two beams will widen the fringes.

If we reduce the angle to zero, the result will be a single large fringe covering the entire Airy spot. By careful design, you can make this a light or a dark fringe and have the Airy spot turn on or off completely. Remembering that this Airy spot is the image of a star, we can use this to completely negate the starlight in our image. Why would we want to do this? After all, we have gone to so much trouble to collect the light in the first place, it seems odd to want to get rid of it right at the end.

The secret lies in the fact that we have merely aligned the interferometer in such a way as to give zero fringe (called a 'null') for that particular star. However, for an object just off to the side, the light may combine to give a bright fringe – so the object remains visible. In this way we might be able to 'see' a dim object even if it lies extremely close to a much brighter one. The light from a star is typically millions or billions of times brighter than that from even a large orbiting planet, so normally the starlight would completely dominate the image. With nulling interferometry, there is some hope that the starlight could be cancelled out enough to image the nearby planet. Well, at least this is the idea. This sort of interferometry requires precise control of the light beams and extremely well-constructed telescopes. It is in making these telescopes that time, effort and money start to come into the equation.

Chapter 8

So you want to build an observatory?

We have read about how to make a telescope and how to ensure that it is free of aberrations. This gives you an idea of the requirements, but it is mostly a theoretical exercise which tells little of how to go about, say, making a world-class observatory. There is more to building an astronomical observatory than simply bolting a couple of big mirrors together, attaching a camera and pointing it at some nebula to take happy snaps. Without a doubt, the two most important factors to consider from the outset are the mirror and the site. The mirror will affect the types of observations made and how much light will be collected, while the site determines how well you will be able to see objects.

Making a mirror

Making a large mirror is a costly and time-consuming process and more than just a little stressful for those involved. This is not simply a result of the high degree of precision involved, but also because the slightest mistake could result in a defect, flaw or damage which would ruin the entire project, wasting years of effort and millions of dollars. As you will see, it is definitely not a job for the faint of heart.

The first thing we will need is glass. Normal glass used for everyday applications, such as windows and bathroom mirrors, is soda-lime glass. Although used in the past for telescope mirrors, soda-lime is rarely used today as it expands greatly under small temperature changes. We have a measure of this property, called the coefficient of thermal expansion (CTE). In a telescope which is operated even over a small range of thermal conditions, we cannot allow such expansions and contractions of the primary. Even a small CTE over a mirror with a diameter of 8 m will result in a dramatic change of shape over a few degrees Celsius. With this in mind, glass manufacturers have developed speciality materials which have negligible CTEs.

The first major developments in this field were Pyrex and fused silica, used for the Hale and ESO 3.6 m telescopes, respectively. Further research has a resulted in the development of two better materials: Zerodur and Ultra-Low Expansion (ULE) glass. These two have a CTE that is essentially zero (hence the names), in spite of having relatively poor thermal conductivities. This is important because we want to have everything in our observatory at the same temperature as the outside world. In the daytime with the dome closed, the telescope will heat up above ambient outside conditions. When we open the dome up at night, we want the mirror to cool as quickly as possible, and a high thermal conductivity makes this possible.

The two speciality glasses mentioned here do not come cheap, and can end up being a significant portion of the cost of the mirror. This is because the make-up of the final block of glass must be completely homogeneous. The slightest variations in density from one part of the mirror to the next (such as cracks or air bubbles) will result in strain and deformation over time. Once we have bought our glass, we then have to cast the blocks together to form one large 'blank'. These days, solid slabs of glass are rarely used for large telescopes. Instead, they tend to be made 'lightweighted', where a thin facesheet is supported by a structure on the back which often looks like a honeycomb. The ribs in this honeycomb provide stiffness without all the mass of a thick, solid mirror. Still larger telescopes can be made of smaller mirror segments which are then mounted together to act as a single large mirror. In either case, the mirror has to be cast and gradually cooled to room temperature. This has to be done very slowly to let all the thermal stresses dissipate and to avoid cracks forming. Often the casting process can take weeks to complete.

Followed those steps? Congratulations, you are now in possession of what is essentially a big block of glass which can be fashioned into a mirror by grinding, polishing and figuring. The shorter the focal length of the primary, the more material must be removed, and the longer the process will take. These days, primaries are made very highly curved to shorten the length of the telescope. While it increases the cost of the optics, it makes the dome much smaller, which greatly reduces costs and seeing issues. Let's say we have an 8-m mirror with a 16-m focal length. This will mean that material will have to be removed to a depth of 25 cm in the centre, amounting to the removal of over 10 tonnes of glass.

To begin with, the bulk of the material is removed by computer-controlled diamond cutting to create the correct aspheric shape (e.g. paraboloid or hyperboloid). The grinding process is then begun by spinning a disk in light contact with the mirror blank with a slurry of abrasive, gritty material suspended in water. Initially the grit is very coarse, but as the surface gets closer to that desired, finer and finer grains are used which scour the surface more and more gently. The grinding process removes still more material and gets the surface good to within a few micrometres. The next step is polishing, which requires an extremely fine grit and a polishing disk (called a lap). Gentle rubbing of the glass continues until the mirror is smooth to about a few nanometres and there are no significant scratches anywhere. The last step is to figure the mirror using a lap which can be moved and deformed under computer control. All in all, the process takes many months or even years, not due so much to difficulties in removing the material, but rather because of the extensive testing required along the way.

It is not simply enough to tell a computer to 'make this shape', press the Enter key and *voilà!*: a high-precision surface. After the initial figuring process, the mirror has to be tested to see if the shape is exactly what we want or not. At this stage, it may be discovered that portions of the mirror have too much or too little curvature, so the mirror must be re-figured to correct for these errors. Often, this testing and refiguring must take place more than once and is tedious and time consuming. It also has to be done very carefully or the mirror could be damaged or, as in the case of the Hubble Space Telescope, incorrectly figured. Breaking such a mirror will give you more than seven years of bad luck – it will cost you millions of dollars and all your friends.

So how do you test a mirror? There are several methods with various merits and limitations. I'll spare you the laundry list and simply give a basic run-down of the process. To begin with, we need to light the surface with a point source of light. Generally we choose a place like the centre of curvature where we can expect the reflected light to come back to a point again. By analysing the wavefront reflected from the mirror, we can see whether it has the precise shape we want. One test is to use interferometry to generate a contour map of the surface distortions, as shown in Figure 7.5. These tests have to be made using extremely high-quality optics and in climate-controlled labs to reduce atmospheric effects. Moreover, each time errors are found, the mirror will have to be tested again after corrective polishing to see if any other errors have arisen. Needless to say, this process is fraught with potential error, so this is where a lot of the expense comes in. But if you can stick it out you should now have a polished piece of glass with a high-quality figure.

At this stage the mirror is still just a block of glass with a shiny concave surface. Glass typically has a reflectivity of a mere 4% in the visible, so we have to apply a metallic coating in order to increase the reflectivity. The mirrors you come across in everyday life are coated by 'painting' a reflective chemical on the rear of the glass. This sort of chemical coating tarnishes easily with exposure to the air. Putting it on the back is not just to prevent casual scratching, but also so that the glass protects the part of the coating that we see. Such an arrangement would not be suitable for a telescope mirror as we would always get a dim 'ghost' image produced by reflection off the front surface. The solution is to deposit a layer of metal directly onto the front surface. This is done by placing the mirror in a large vacuum chamber where a filament of metal is vaporised by a high electrical current. Where the mist of metal atoms strikes the mirror, they condense to form a thin layer, typically less than a micron thick.

Aluminium is generally the preferred metal, with a few telescopes opting for silver or gold, which are more reflective in the infrared. The reflectivity of a front-surface aluminium coating can be over 95% in the visible compared with a typical value of 80% for common back-surface mirrors. However, front-surface mirrors are extremely difficult to take care of. Often they are given an equally thin protective overcoat of transparent magnesium fluoride, but if they are bare, they have to be carefully cleaned and maintained.

Telescope mirrors should *never* be touched, as this will leave an oily residue which will scatter light, cause chemical degradation of the surface and lower reflectivity. As a result, cleaning a front-surface mirror is a daunting prospect. It is not simply a case of spraying on window cleaner and wiping it with a rag. For small mirrors you can use high-purity acetone or alcohols which are washed off without leaving any residue. I once met a Russian physicist who told me how they used to do it in their lab. They would take a swig of vodka, breathe the fumes over their optics and gently wipe it with a thin lens cloth. Works like a charm. But liquids are not always an option for large telescope mirrors. The fumes would be overwhelming in the quantities required, and generally all we need to do is to get rid of layers of dust that build up over time. Instead of using liquids, many observatories use carbon dioxide similar to that from a fire extinguisher. The CO_2 snow is sprayed on the mirrors and as it gently slides off, it takes the dust with it. Even with this approach, the mirrors will eventually degrade through oxidation, so they have to be recoated every couple of years. Most larger facilities have their own large vacuum coating chambers located in the observatory for this express purpose.

By now it should be clear why making and maintaining the primary mirror of a telescope is an expensive affair. Many hours of work go into ensuring high manufacturing standards as well as long-term precision operation. And as yet we have only considered the primary. Making the secondary is generally a much simpler task, due to its smaller size. However, because the mirror is convex, the testing procedure is often much more complicated. Instead of a simple test procedure where a point of light is refocused back to a point of light, we have to generate a large diameter focused beam to reflect off the surface. Again, this test is only as good as the worst test optic, so the larger optic has to be made to a high precision itself. Even so, the cost of the secondary is a relatively minor expense compared with that of the primary.

Site selection

When telescopes were first constructed, the choice of a site was given a minimal amount of thought. Even when it was discovered that atmospheric seeing was imposing limits on resolution, little was done to find the optimum site for an observatory. For example, the 2.5-m Hooker telescope is located on top of Mt Wilson just outside Los Angeles. Yes, it is at the top of a mountain, and yes, cloud cover and precipitation are generally low, leaving many nights available for astronomy. But the choice of a mountain close to Los Angeles was mostly a case of convenience. By the time the 5-m Hale telescope was constructed some 30 years later, light pollution from Los Angeles was affecting observations, so a new site was chosen on Mt Palomar, halfway between Los Angeles and San Diego to the south. Still, this observatory is at a mere 1900-m altitude, and the reasons for choosing the site still had more to do with proximity to a city than optimising seeing. Also, little attention was paid to ground vibrations, which are a significant problem in California.

Until recently, these problems were considered minor issues, because the telescope would not be working at the diffraction limit anyway. Atmospheric turbulence would always be the ultimate limit in resolution, so who cared? These days, with some of the advanced techniques to be described later, telescopes can approach diffraction-limited performance, so the selection of the site becomes vitally important. Here are some of the major considerations.

Seeing

We want a site with the least atmospheric turbulence which, in turn, gives the best seeing. In general, we would search for a mountain-top to get above as much of the intervening air as possible. At first glance it would seem that somewhere in Nepal would be the best location in the world, but altitude is not everything. There is also location to be considered. In Nepal, the prevailing winds have travelled over many hundreds of kilometres of land. Differential heating effects from land mean that the motion of the air becomes greatly unsettled by the time it reaches the Himalayas. A more ideal location is on a mountain where the prevailing winds come off a large stretch of ocean – hence the preference for Hawaii and Chile. The water is more uniform in temperature and smoother, which results in much more laminar airflow.

Weather

A mountain top is a good solution for seeing, and in many cases also works for getting the observatory above as much of the cloud and bad weather as possible. This in turn increases the number of nights which can be used for actual observations, and the quality of the observations themselves. High altitudes also have thinner and cooler air with lower humidity, which is better for instruments and for observing at infrared wavelengths.

Light pollution

Heavily populated cities beam a lot of light into the night sky. Street lights, car headlights and personal/industrial lighting can either directly illuminate or reflect into the sky, causing air molecules and dust to glow brightly. Even a small amount of light pollution can reduce the effectiveness of observations of dim objects. Observatories should be located as far from major population centres as possible. Historical considerations of logistics are less of a concern today, since telescopes can be operated remotely and astronomers can even make observations from the other side of the world.[15]

Ground stability

A telescope is a heavy object which must be steered with high pointing precision. In order to maintain a good pointing stability, a solid foundation is required, along with minimal ground vibration. With this in mind, it is a good idea to avoid any location where there may be significant geological activity which would shake the ground and blur images.

With all these factors to consider, a great deal of effort goes into selecting an appropriate location for a multi-million dollar facility in order to optimise scientific value. Nowadays, a site will be tested for seeing conditions over several years, and decades of weather records will be analysed before a choice is made. Even when all the physical testing has been completed, there are environmental and even local issues to be considered. Many of the best sites have issues with local ownership, religious significance and environmental impact that are not to be taken lightly. For example, the peaks of both Mauna Kea and Haleakala are sacred sites for the Hawaiians, who have been very generous in permitting scientists to clutter up the place with massive white buildings. To put it in perspective, one should imagine how Catholics would feel about astronomers turning the Vatican into the site of an observatory – no matter how inviting the dome of St Peter's Basilica may appear. In the case of Mauna Kea, the compromise has been to set restrictions on increasing the size and number of the facilities, so these days it is quite difficult to add even a tiny dome on the site. This continues to be the cause of legal battles for the Keck observatory trying to install a few small (1-m diameter) outrigger telescopes to be interferometrically combined with the two bigger telescopes.

A different problem was faced by the European Southern Observatory (ESO) in Cerro Paranal in Chile. In 1988, ESO secured use of the land from the Pinochet government as the site for their four VLT

8.2-m telescopes. Soon after the mountain top had been levelled, a local family claiming ownership of the site began legal proceedings to halt construction. After three years of legal battles, the issue was resolved with $15 million in compensation paid to the family. Meanwhile, two problems arose with the installation of the Large Binocular Telescope on Mt Graham in Arizona. The chosen site was sacred to Apache Indians and it also turned out to be part of the habitat of an endangered red squirrel. Although these issues were eventually resolved, they highlight some of the unexpected problems behind site selection.

So where is the best ground-based site in the world? Hawaii and Chile are roughly tied for this honour, so not surprisingly, this is where the premier telescopes are located. However, recently published survey results from 'Dome C' in Antarctica (75° S, 123° E) show that it may provide seeing twice as good as either of those sites. According to our criteria, of course, this should not come as too big a surprise. Antarctica is the highest continent on Earth. Dome C itself is at an altitude of 3250 m and the ground is very cold, which reduces turbulence. Even better, the atmosphere with its low temperature and humidity make it an ideal location for infrared observations. The lack of dust or light pollution is also particularly appealing for astronomers. Overall a great choice – the only problem being that it is very remote (and cold), which increases operational and manning costs. However, it may end up being a hard sell for next-generation large telescopes merely because astronomers would prefer to make their trips to Hawaii instead.

Mechanical engineering

At this stage we have made our mirror and chosen a site. We now have to carefully mount the optics into a rigid structure, and then work out a way to precisely steer them to track objects across the sky. Oh, and they'll also need to be housed in a protective structure with a big hole in the top so we can see out. Sounds easy, right? For the most part, these engineering issues are fairly straightforward, but we'll go over some of the design issues just in case you have the desire to build an 8-metre-class facility in your own backyard.

To begin with, you have to choose your telescope mechanical configuration. This is not to be confused with the different optical designs we discussed in Chapters 3 and 5. Here we are talking about how we get to the focus, which in turn affects the location of instruments as well as mounting and tracking issues. The most prevalent configurations are shown in Figure 8.1. The prime focus is sometimes used for imaging but more often for photometry and lidar (which we will discuss in Chapter 11). The Cassegrain focus is more common for older telescopes and small amateur devices, but is not always practical as there is a limit to the size of the instruments that can be attached to the rear of the primary. The most familiar configuration for large modern telescopes is the Nasmyth focus as the image can be brought out to platforms at the side of the telescope. On these Nasmyth platforms we can put very complex and heavy instruments. The last focal arrangement is the Coudé path, which involves transferring the light to a stationary location. The Coudé path is mainly used for specialty applications where the reduced field of view is acceptable.

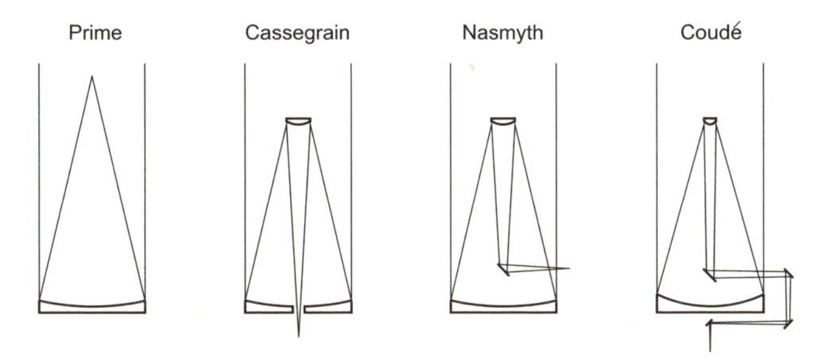

Figure 8.1: Telescope configurations. The light in a telescope can be focused in front of the primary (prime focus), behind the primary (Cassegrain), off to one side (Nasmyth) or some distance to the rear via transport optics (Coudé).

Most major telescopes built before 1950, and in fact all those you will still buy in a shop today, will have a solid tube enclosing the two mirrors. Why? Because for typical operation, such amateur telescopes need to exclude dust and stray light from getting into the final detector (either the camera or your eye). With large telescopes located at the top of a mountain in a darkened dome, this is less of a problem and is eliminated using baffles. In fact, these days we go out of our way to avoid covering our optics. Inside an enclosed tube, air is trapped and will generally have a different temperature from the air outside. This thermal differential causes a slight amount of turbulence. The answer, then, is to use an open truss structure to support the optics, with no trapped air. The telescope will have the primary and secondary mirrors located at either end, with the support ring at the centre of mass somewhere in the middle. The support ring is then mounted to a yoke which sits in the tracking mount.

Mounting and steering the telescope requires some precision engineering in itself. After all, there's no point in having a high-quality camera trying to take pictures of something if it is shaking all over the place. The mount should be as stable as possible and capable of precision tracking of celestial objects. Until the advent of computers, telescopes were steered on equatorial mounts. This simple method requires tilting one axis of the telescope precisely towards the polar axis. Tracking of any object in the sky can then be achieved simply by turning about this axis at sidereal rates (i.e. one complete rotation in 23 hours, 56 minutes and 4.1 seconds). Different objects can be brought into view by changing the angle of declination, but always the rate of rotation is constant. Building an equatorial mount is relatively simple, but the downside is that the massive telescope is being suspended out at the end of a long arm. For extremely large telescopes, this is not a very stable arrangement. The solution came with the advent of the altitude-azimuth (alt-az) mount. A comparison of the two types can be seen in Figure 8.2.

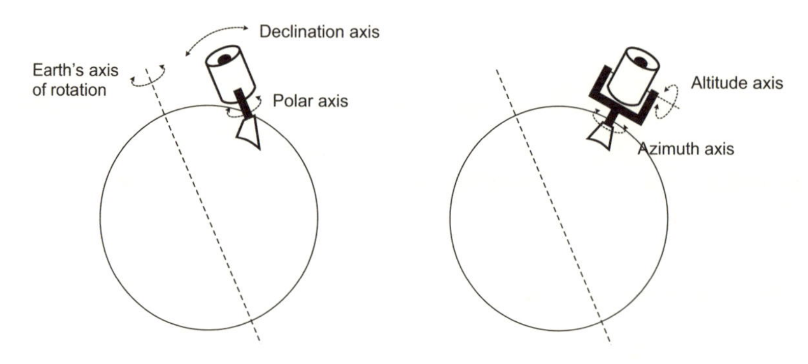

Figure 8.2: An Equatorial mount has one axis aligned to the Earth's axis of rotation. The Altitude-Azimuth mount has one axis aligned to local zenith (straight up) and the other at right angles to it.

The alt-az mount (sometimes called the elevation-azimuth) is a much more stable configuration for a telescope, with the telescope centre of mass always located directly over the ground support. The tracking is achieved by careful rotation about two axes at right angles to each other. Altitude is the height above the horizon (given in degrees, minutes and seconds), and azimuth is aligned much like a compass direction (but is given in hours, minutes and seconds). The downside to the alt-az motion is that this is not the way stars move across the sky, so the rotation rate of each axis must be constantly varied for each and every point in the sky. For ultra-precise tracking, this sort of complex motion of a heavy instrument has only become possible with the advent of computer control. These days, virtually every large telescope uses the alt-az mount while most backyard telescopes continue to use the simpler polar mount.

Having said this, there is of course a class of telescopes which forms a notable exception. The Hobby-Eberly Telescope (HET) in Texas and the newly commissioned Southern African Large Telescope (SALT) in South Africa have taken a slightly different approach to combating the mounting costs, so to speak. The solution is to reduce the complexity and only allow the telescope to move in a single axis. In the case of both the SALT and the HET, the altitude/elevation is fixed, but they can still be moved in azimuth to reach much of the sky over the period of a year. These telescopes don't track in the conventional sense – instead they move their secondaries to change the effective pointing direction. This reduces the effective collecting area of the primary, but by most estimates, the construction costs are almost halved. Both of these telescopes are over 10 m in diameter, and while they do have less flexibility in what they can point at, they allow astronomers to get massive light-collecting power for a fraction of the costs of other 10-m facilities.

Once you have decided how to move the telescope, you then have to make sure you do so smoothly and accurately. Given that large telescopes weigh in at thousands of kilograms, this is no easy matter. We have already mentioned how time is actually based on the motion of the distant celestial sphere. Thus our tracking of objects requires very accurate time-keeping. Think of it this way – the stars will move at roughly 15 degrees per hour overhead, which equates to 15 arcseconds per second. Now, if a 10-m telescope were able to operate at the diffraction limit, it could resolve around 0.01 arcseconds. This is the distance a star would move in 67 milliseconds, so our timing has to be at least this good. In fact, astronomers often use timing measurements much better than this.

For each activity a 'time-stamp' is attached in the form of a Julian date (JD). It doesn't make sense for astronomers to use our calendrical method of recording time as this makes it difficult to set a timeline for measurements over months or even years. Quickly – tell me the length of time between 8:05:01 am on 20 September 2005 and 4:56:32 pm on 9 February 1970. Tricky, isn't it? Instead, astronomers use a Julian date which was set as beginning at noon UT[16] on Monday, 1 January 4713 BCE. Since then, every day has added one more to the number, while parts of the day are reduced to decimals. So at 6:10:44 am (UT) on 1 January 2007 the time will be 2454101.75745 JD. Simple. Now you only have to know where to find your object. But objects don't move as simply as you might at first expect. Several corrections need to be applied to know where an object is and how it moves in the course of a night.

First we need to know where the object is. While most objects have had their co-ordinates catalogued, there is still the problem of motion over time. The slow motion of the Earth's axis of rotation due to precession (Chapter 1) is one such effect. But there can also be appreciable proper motion which is the apparent tangential motion of objects with respect to the Earth. Generally this is something which is only important in the case of objects quite close to the Earth (planets, asteroids, nearby stars, etc). For this reason, we define an object's location according to a particular epoch. Another effect is stellar aberration, which is the apparent motion of an object according to its latitude. The effect is due to the finite speed of light, which causes the object to lead or lag its average position due to the motion of the Earth. This aberration can shift the location of a star by up to 20 arcseconds.

Perhaps the largest source of stellar motion is the result of the refractive index of the Earth's atmosphere. When an object is directly overhead, the light travels through the atmosphere without refracting. Nearer to the horizon, the light will bend as it passes through the atmosphere to appear higher in the sky than it actually is. In effect, this 'slows' the setting of a star compared with the expected motion and the angular velocity will vary with altitude/elevation above the horizon. Worse still, the amount of refraction varies with the height of the observatory above sea level (as the amount of air the light passes through will be different). All in all, we have a lot of calculations to worry about simply to follow an object in the sky.

At this stage we have the telescope mounted and tracking so we now need to house it. Most people have the same idea when they think of observatory domes – big white mushrooms sitting on top of a mountain. The reason they are white is simply to reflect the most sunlight during the day to reduce heating of the air within the dome. Any sources of heating are to be avoided as this hot air will mix with the cold air at night causing local turbulence called 'dome seeing'. But why are domes hemispherical? In the past, it was thought that this was the best shape to minimise turbulence of the airflow around the structure. More recent work has shown that perhaps other shapes are suitable, which is why the VLT and LBT telescopes are in box-like 'domes'. By far the most impressive dome is the enclosure for each of the twin 8-m Gemini telescopes. Here the engineers incorporated large vents around the entire dome which can be opened at sunset to allow warm air to escape and permit free flow of air through the structure at night.

Figure 8.3: Gemini North in Mauna Kea at sunset. The central dome slit is open, along with the vents around the edge. Courtesy: Neelon Crawford, Gemini Observatory/AURA.

At this point we are getting close to the moment when we commission our new instrument by taking the initial image. Astronomers call this 'first light', and there is a great deal of nervous anticipation among those standing around the computer screens waiting for the first image to be displayed to show the fruits of years of labour. I haven't discussed much about the control of the telescope, but as we have seen, astronomers never look through an eyepiece any more. In fact, warm bodies are never welcome inside the cool dome at night. Since the entire system is under remote control, the telescope and instruments are generally operated from buildings elsewhere on the mountain. This idea has been taken a step further with the 2-m Liverpool Telescope. This instrument sits on Haleakala in Maui, but is remotely controlled by schoolchildren in England. The fact that the telescope is on the other side of the planet means that it can be operated during daytime classes to view the nighttime skies over Hawaii. Taking remote operation to the fullest extent, we don't even need our telescope to be on Earth at all – which brings us to the Hubble Space Telescope.

Chapter 9

The Hubble Space Telescope

It is the morning (astronaut time) of 25 April 1990, and 615 km above the Earth, a new telescope is being removed from the payload bay of the Space Shuttle *Discovery*. This represents the culmination of over two billion dollars spent and 20 years' planning and development by thousands of people. Throughout the following weeks, the telescope is carefully taken through a large list of procedures designed to check out the systems and ensure that everything is working according to the design specifications. As always happens during this phase of any mission, several glitches are discovered. These are either solved by sending up new software commands from the ground or avoided altogether by altering the telescope's operational procedures. So when the first telescope image is snapped on 30 May 1990, it is only a minor cause for concern that it appears somewhat fuzzy, because this is initially dismissed as a minor problem with the focus. Since the telescope mirrors are designed to be moved to adjust the focus, it is expected that the correction can be applied simply by optimising the mirror positions. Alternatively, the blurring might be due to some residual moisture which will escape into space over the next few days. Either way, it's mostly good news, and nothing much to be concerned about.

Several days later, engineers at the NASA Goddard Space Flight Center (GFSC) and the Space Telescope Science Institute (STScI) in Baltimore were feeling very different. In place of a minor twinge of annoyance was the ominous feeling that something was very wrong. No one wanted to be the first to say what they were all thinking. Quite simply, the telescope was broken – one of the mirrors was the wrong shape and there was a massive amount of spherical aberration. As the truth began to sink in, the project scientists and engineers grimly contemplated what this meant for the future of the space telescope. One or two also probably gave thought to the many years of planning, design, testing, retesting, documentation and intricate management of the project. It was impossible to fathom just how this could have happened. In fact, it would take several months to conclusively identify the cause of the error and to find that the billion dollar project was undone by a few flecks of paint. Just as Benjamin Franklin had penned some two hundred years earlier: 'But for the want of a nail ... the kingdom was lost.'

The story of the Hubble Space Telescope (HST) reads like the classic Hollywood Cinderella script. Beginning with a few dedicated supporters, the idea slowly gains momentum and builds with relentless anticipation. Suddenly the time comes for the promise to become a reality. The world holds its breath. Then comes the failure ... and the fall from grace. Popular opinion turns in an instant and the hero becomes the goat. The champion becomes the underdog. Then begins the comeback and the return to undreamt-of successes. Heady stuff indeed. Yes, this story has it all – including the massive budget befitting a Hollywood blockbuster. The producers would probably like to do a little creative editing of the records to give a personal angle. Perhaps a battle between the managers/publicists dressed in black suits, and the scientists/engineers dressed in clean white lab coats. As we will see, though, this story is not so simple.

Most people would agree that the tale should begin with Lyman Spitzer, a scientist working for the Rand Corporation. In 1946, Spitzer published a report detailing the advantages and the technological possibility of using the wartime rocket technology newly acquired from Germany to put an optical telescope into orbit around the Earth. The idea of a space-based telescope had been considered by many over the years. Long before the Soviets launched Sputnik, the advantages of such an instrument were well documented. The first and most obvious advantage is that in space there is no atmospheric turbulence to blur images of distant objects. We have discussed this before, but perhaps the graph in Figure 9.1 tells the story best. Since Galileo's time, the effective resolution of telescopes had not improved much because of the limits imposed by the atmosphere. Putting a 2.4-m diameter telescope in Earth orbit would mean a tenfold jump in resolving power similar to that which occurred with the invention of the telescope itself. Given the revelations which resulted from Galileo's use of this new gadget to observe the heavens, it is easy to imagine that equally important and unforeseen discoveries would be made using a telescope in space.

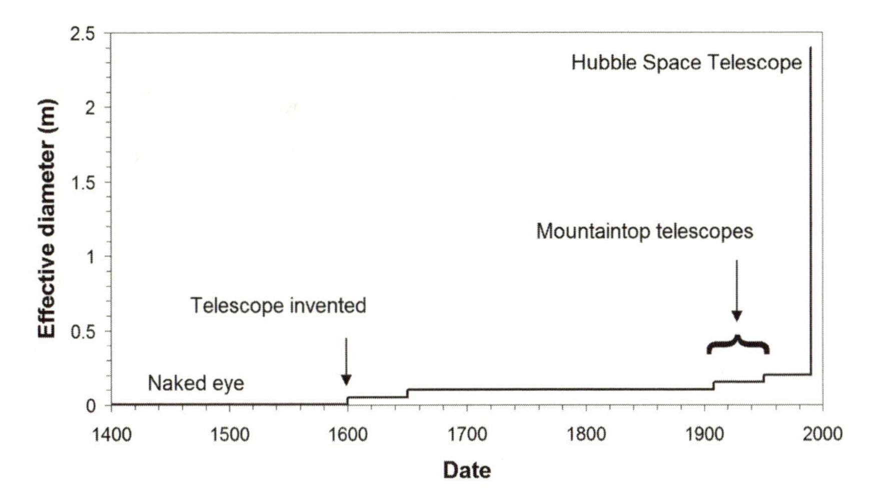

Figure 9.1: Historical increases in optical telescope diameter have increased light gathering but have contributed little in improved resolution, because of problems with seeing. A space telescope results in a leap in resolution not seen since the invention of the telescope. This graph does not include the advent of adaptive optics, which will be discussed in the next chapter.

Another advantage of a space-based instrument would be the ability to observe objects from the near-infrared all the way down to ultraviolet wavelengths. If the primary is figured and polished to a sufficiently high-quality surface, imaging would be possible even at wavelengths as short as 100 nm. Not only is this a fourfold improvement in resolution over optimal visible imaging, but these wavelengths are not even accessible to ground-based telescopes. This is due to the fact that for wavelengths shorter than about 300 nm, the atmosphere becomes essentially opaque. Figure 9.2 shows a plot of the atmospheric transmission for wavelengths from 300 to 900 nanometres.

The last major benefit of a space telescope is the ability to observe very dim objects. An orbital location makes it possible to observe some portions of the sky for uninterrupted periods as long as weeks or months, if so desired. For these areas, there is no daytime sky or horizon which would cut down observation time to a mere eight to ten hours a day, so we can build up images of very dim objects.[17] Added to this is the absence of sky glow, both artificial and natural. Even in the darkest sites on Earth, the sky will have an overall glow due to its internal heat, which will increase the noise levels of long-term observations of dim objects. In space there is no sky glow, so we can see dimmer objects by observing them over longer periods of time.

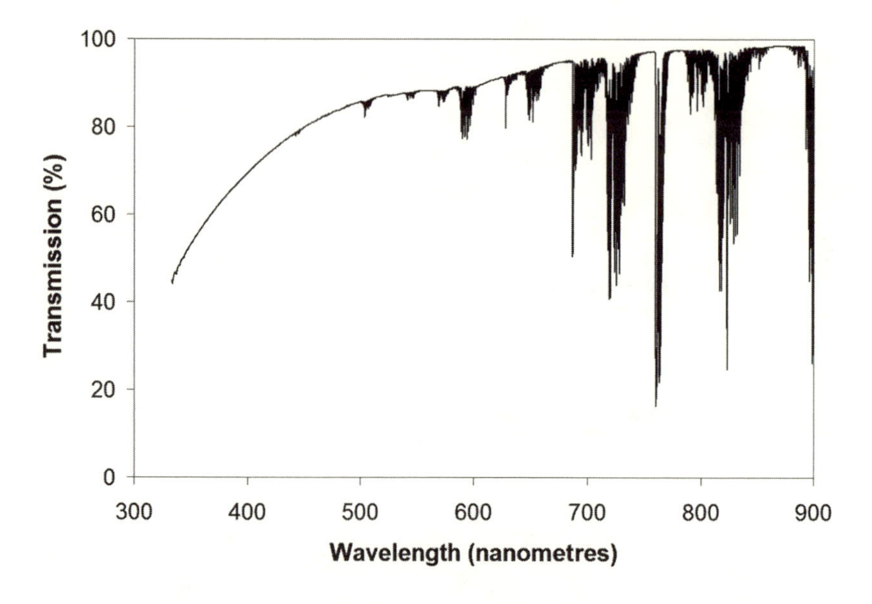

Figure 9.2: This plot shows how the transmission of light through the atmosphere varies with wavelength in micrometres.

These phenomena are just a few of the major observational improvements to be expected from a large space telescope. But still – why do it? There is a huge amount of expense in putting even the smallest objects into orbit (around $10,000 per kilogram), so the scientific payoff has to be worth it. In this case, the main reason behind the launching of the space telescope was that using it would, it was hoped, lead to an answer to two of the big questions in astrophysics: 'How old is the universe?' and 'What is the ultimate fate of the universe?' To better understand how a space telescope could help us in these endeavours, we first need a little background.

At the turn of the twentieth century, it was widely believed that the universe was infinite in both age and dimension (or time and space, if you'd prefer) and unchanging. This so-called 'steady-state' theory had no real basis in observation, but was simply what most people felt was right. A big question yet to be answered, however, was the distribution of the objects within the universe. Some people believed that absolutely everything we see 'out there' was all-inclusive to the Milky Way galaxy. Others believed that the Milky Way was simply one of a number of island galaxies which existed as separate entities. Part of the problem to be addressed was the nature and distance of several different types of fuzzy (nebulous) objects. Eventually some were found to be glowing clouds of luminous gas (now called nebulae), which formed part of our own galaxy. Meanwhile others were found to consist of individual stars and merely appeared to be cloudy blobs, much as the Milky Way itself looks when viewed with the naked eye. This seemed to suggest that the second class of objects were other galaxies, which existed as separate entities to the Milky Way. But the scale was still a mystery.

One piece of evidence was gained by an American astronomer, Vesto Slipher. In 1913, he used Doppler-shift measurements of Andromeda to show that it was moving at 300 km/s towards us! Clearly it must be a long distance away or it would noticeably change position and size in the sky over time. This hinted at a large scale for the universe, but if you don't know how large a galaxy is, then how can you measure its distance? The answer was found by two Harvard astronomers: William Pickering and his research associate, Henrietta Swan Leavitt. In 1908, Leavitt began observing stars in the Large and Small Magellanic Clouds which had a particular property in that they varied in brightness over a period of time. This class of stars is known as Cepheid variables or simply Cepheids for short. The name derives from the first such star discovered by John Goodricke in 1784, called Delta Cephei.[18] Leavitt showed that there was a relationship between the apparent magnitude of Cepheids and the period of their brightness variation. Over the next four years, Leavitt refined the period-luminosity relationship and showed how it could be used to nail down the distance to the Large Magellanic Cloud (LMC) and the Small Magellanic Cloud (SMC).

The idea was to use the Cepheid variables as 'standard candles'. If we know how bright a star actually is (absolute magnitude), and compare this with how bright it appears to be (apparent magnitude), we can calculate how far away it is. This is like finding the distance to a candle by measuring how bright it looks from a distance compared with the brightness of a similar candle nearby. The key here is the knowledge of what a nearby candle (or star in this case) looks like. First you need to find an accurate distance to a nearby Cepheid, so that an absolute brightness can be associated with the period. From there, we can set the scale of brightness-luminosity to tell us exactly how far away a star is by how its brightness varies. Then the scale of the universe could be set simply by observing Cepheids in other galaxies. This was precisely what Edwin Hubble set out to do.

Edwin Hubble was born in Missouri in 1889. A Rhodes Scholar and practising lawyer, he changed fields to astronomy because, as he said: '... I knew that even if I were second-rate or third rate, it was astronomy that mattered.' In 1919 Hubble began working at Mt Wilson, and by 1923 he was using the 2.4-m Hooker telescope to search for Cepheids in the Andromeda galaxy (M31). From his measurements, he got an estimate for the distance to be around 900,000 light-years (now known to be closer to 2.9 million light-years). These (and subsequent) measurements made it fairly clear that the M31 galaxy did in fact lie well outside our own galaxy and was a galaxy in its own right. Its apparent size in our night sky also indicated that it was huge – containing billions of stars. Over the next five years, Hubble extended his measurements and used alternative techniques to measure distances to cosmic objects up to 500 million light-years. This was significant as it indicated that the universe was much larger than anyone had previously thought. If that were not enough, Hubble then turned his attention to the job of measuring the motion of these galaxies, and made an even more surprising discovery.

In 1917, Vesto Slipher had shown that but for a few exceptions (such as Andromeda), all galaxies seem to moving away from us, as evidenced by the redshift of their spectral lines. The steady-state model for the universe had it that all galaxies should be evenly distributed in space with no significant motion between them – rather like chocolate chips in a tub of ice-cream. This was the opinion of most other scientists, including Newton and Einstein. In 1929 Hubble announced the results of observations made in collaboration with his assistant Milton Humason. By combining measurements of distance with redshift, Hubble was able to show that the rate of recession increases with distance. The implications of this finding were huge. Not only did this mean that the universe was expanding, but it also implied that at some stage in the past the universe had been a very crowded place.

It is important to emphasise that this 'Cosmological Redshift' comes from the expansion of space-time itself. It is not to be confused with the lesser inherent (proper) motion of the galaxies themselves, nor is it the result of the creation of new matter or 'space'. Imagine it this way: blow up a balloon and mark two points in pen near each other on the balloon's surface. Inflate the balloon some more, and the two points will increase their separation. Furthermore, the more distant the two points, the larger their rate of separation will be. This being so, it means that we can imagine a time in the past when the universe was much more compact, and that the birth of all the matter it contains occurred in what is now known as the Big Bang. In order to know when this happened, we need to precisely characterise the relationship between the rate of expansion and distance (see Appendix B for more details). This is given by Hubble's Law, and the critical number in this calculation is the Hubble Constant (H_o). In Hubble's original paper, the value of this constant was calculated to be around 500 kilometres per second per megaparsec, which gave an age for the universe of almost 2 billion years. Later measurements showed that this value was some way off the mark, and subsequent observations have been aimed at finding out exactly what it is.

At the time of the launch of the space telescope, the value of the Hubble Constant had been refined to be somewhere between 50 and 100 kilometres per second per megaparsec. This implied an age for the universe of between 10 and 20 billion years. In order to pin it down with more precision, it was necessary to extend the distance measurements out to more remote galaxies. To do so, astronomers needed a larger and more sensitive telescope to image dimmer galaxies. The problem with the measurements was not with the redshift values, which are relatively simple to make with a high degree of accuracy. It is distance that is much more difficult to measure because of limits on imaging far-flung objects and because of the many different assumptions used in the calculations. The Hubble Space Telescope was aimed at trying to improve distance measurements by allowing astronomers to 'see' Cepheid variables in still more distant galaxies. But even Cepheids are only bright enough for measuring medium-range distances. To measure out to larger distances, much brighter standard candles such as certain supernovae are used instead. Supernovae are extremely bright events, so it should be possible to observe them occurring in galaxies many times more distant than those in which Cepheids could be observed.

With the major emphasis on determining the size and age of the universe, a space telescope would prove invaluable. But astronomers could also use it to study planets in our Solar System, look for clouds of cometary matter around the edges of our Solar System, image distant star-forming regions or help determine the physical make-up of the universe. Observations into the deep cosmos would also make it possible to determine the way the universe and all the objects in it have evolved since the Big Bang. These are just a few of the many reasons for building the telescope, most of which were repeated many times in years of mission-funding meetings. Then in 1974, NASA officially announced its support for a space telescope project.

It is often incorrectly claimed that the Hubble Space Telescope is the largest telescope which could fit into the payload of the Space Shuttle. In fact, the space telescope was originally planned with a 3-m diameter mirror, but this diameter was later reduced to 2.4 m for reasons of cost. Two companies competed for the mirror fabrication – Eastman Kodak and Perkin-Elmer/Goodrich.[19] Both companies fabricated blanks, with the eventual contract going to Perkin-Elmer for a bid of $70 million compared with the Kodak bid of $100 million. When all was said and done, the final mirror was actually delivered at a cost of $450 million. This massive cost overrun would become a fairly common problem throughout the course of the telescope project – and still continues in similar projects today. This is not to suggest that the mirror was easy to fabricate or was anything less than a technological marvel.

The Hubble Space Telescope primary is made from Ultra-Low Expansion (ULE) glass. In order to minimise the mass of the mirror, it was constructed from several pieces: a 50-mm face sheet and back plate either side of a 0.25-m-thick rectangular honeycomb of glass cells. The three pieces were fused together in an oven to produce a 900-kg blank. The front surface of this blank then had to be ground and polished down to the desired hyperboloid. Since the mirror was designed to go into space, the blank had to be supported in a way that would simulate a zero-gravity environment during the figuring process. The reason for this is that you don't want a mirror made to one shape on Earth to take a different shape in orbit. To accomplish this, the mirror was supported on a bed of 134 precisely located supports which reduced the strain on the mirror to zero. The grinding took two years to complete, with some 70 kg of glass removed in the process.

The next step is polishing, which in this case was especially difficult. The design specifications were that the surface had to be good to 10 nanometres. This detail was more significant in the case of the space telescope than for any other mirror ever constructed. While an error of as much as a wave can be acceptable (and barely noticeable) for a similar sized large ground-based telescope looking through an aberrating atmosphere, the same could not be said for this mirror. In space, any departure from a perfect surface larger than $\lambda/10$

at a wavelength of 115 nm would noticeably affect the final image quality. It has often been reported that the Hubble mirror was the most perfect mirror ever created. Even without the problems discovered later on, this statement is at best stupid and at worst completely disingenuous. Mirrors much smaller than this have been made with much better surfaces, and other mirrors have been made less perfect but significantly larger. The combination of the two – a high-quality surface and large diameter – had never before been achieved, but there is no unit of measure defining these combined qualities.

The mirror testing was always going to be the tricky part as nothing like this had previously been attempted. During testing the slightest breeze, change in temperature or vibration would result in disturbances ten to a hundred times worse than the error being measured. Of course, this was anticipated in advance and was one of the reasons for the high price tag. For the best results, several different tests were conducted. One was a simple, age-old method called a refractive null, while the other was a more complicated, but potentially more precise method, called a reflective null. The results of many refractive null tests seemed to indicate some error in the shape of the mirror, but the reflective null test checked out okay, indicating that the perfect shape desired had been achieved. Due to time constraints and budget overruns, the discrepancy was not fully investigated, and in the end it was decided that the reflective null was giving the true measure of quality, and the mirror polishing was terminated. The glass surface was coated with aluminium and the final mirror was delivered in 1983 – a total fabrication period of some six years, and an estimated four million work hours. And after the expenditure of all this time, money and effort, a critical error in the mirror shape was overlooked.

Meanwhile, the rest of the telescope was also coming together. The HST is a Richey-Chrétien; the secondary (a hyperbola 0.3 m in diameter) was cast, ground, polished and coated to an even better surface quality than the primary mirror. The two mirrors were then combined in a structure a little over 13 m long and 4 m wide (not including the solar panels) – about the size of a short bus. The secondary mirror sits only about half-way down the length of the tube, which is extended out to block out stray light. Baffles were also included and all the inside surfaces were coated with an ultra-absorbent matte black paint. This further blocks out any light which could potentially strike the mirror from directions other than that in which the telescope was pointed. The light entering the telescope is focused by the primary onto the secondary, then through a hole in the primary to an image plane with a field of view of 28 arcmin. The clever design of the telescope is such that several instruments can share the image plane simultaneously, greatly improving its flexibility to users. The main workhorse was the wide-field and planetary camera, but there was also a faint-object camera (FOC), faint-object spectrograph (FOS), high-resolution spectrograph (HRS) and finally a high-speed photometer (HSP). The layout of these cameras is shown in Figure 9.3.

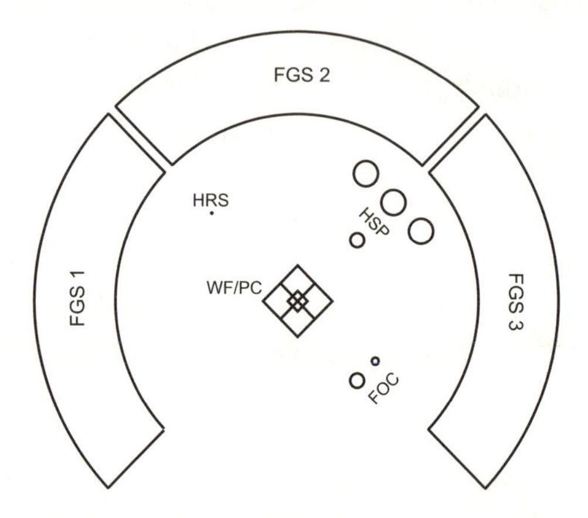

Figure 9.3: The Hubble Space Telescope image plane. The fine-guidance sensors (FGS) lock onto guide stars in each of the three outer segments. The Wide-Field and Planetary Camera (WF/PC) and other image instruments are shown in the central field of view.

The rings around the edge of the field of view are the areas in which the fine-guidance sensors (star trackers) operate. On the ground, a telescope can be rigidly attached to the ground and guided quite easily. The location of the instrument can be known to a few millimetres, and the pointing and tracking characteristics can be modelled and known to a high degree of precision. A telescope in space which requires precision alignment does not have this luxury, as its position and/or orientation are constantly changing. In order to track an object then, the telescope needs a way of referencing its position at any given time with respect to the stars. This is done using star trackers; three portions of the light coming into the telescope are split off with special sensors designed to search for a particular pattern of stars.

The database of stars used for this purpose was previously compiled using the California-based Palomar and Australia-based Siding Spring telescopes. From the thousands of images collected, the positions of hundreds of thousands of stars were recorded. The observing works like this: when an astronomer is approved to observe a particular object, software is used to analyse the known star field around the object and to look for target guide stars (with magnitudes between 9 and 15). The telescope is slewed into rough position using inertial guidance systems, whereupon the star trackers then look for the expected guide stars. Once they are found, the star trackers are programmed to lock onto these stars. Feedback from the fine-guidance sensors directs the telescope to move in very fine amounts so that the overall pointing accuracy is good to about 3 milliarcseconds. In this way, the object being observed is kept perfectly still throughout the exposure.

Here I should point out that the Hubble Space Telescope does not use rocket or gas propulsion for its movements. For a start, the guiding required is hundreds of times finer than this process would provide. More importantly, these propulsion systems would form a cloud of exhaust gases around the telescope. Since one point in having the telescope in space in the first place was to avoid the absorbing and distorting effects of such gases in front of the telescope, it would be a terrible folly to re-create the problem. Instead, Hubble uses gyroscopic reaction wheels for positioning. Like the toy version, these devices consist of a frictionless, rapidly spinning wheel which will point in the same direction in space. By selectively coupling the telescope to one or more (up to three) of these reaction wheels, the telescope can be slowly turned to any particular orientation. Reaction wheels are a very simple and gentle way of precisely directing the spacecraft. This makes them perfect for typical guiding requirements on Hubble, but not so good for compensating fast-changing vibrations or for major changes in orbit.

By 1986, all the engineering issues had been resolved and the 11,000-kg spacecraft was finished and placed in a specially designed cleanroom awaiting launch. Due to a few delays which resulted from a lower than expected launch rate for the Space Shuttles and then the temporary suspension of flights after the *Challenger* disaster in 1986, the Hubble Space Telescope was not launched until 1990, some seven to eight years later than initially planned. Well, better late than never was the general consensus among astronomers. Some would feel a little differently three months later when the first images showed that the telescope was not performing as designed. Much has already been written on the Hubble Space Telescope and its problems, so I will merely address the main points.

First there were some teething problems with the instrument control. Every 50 minutes or so, as the telescope crossed the Earth's day-night divide (called the terminator) a large jitter would shake the telescope. At best this would simply cause the images to blur, but more often than not it would cause the star trackers to lose lock on their guide stars so the telescope would slew uncontrollably. The problem was soon found to be caused by the rapid heating and cooling of the struts along the solar cell arrays which occurs when crossing between night and day sides of the Earth. The heating stretched the struts enough to cause a fast shudder in the spacecraft which could not be compensated for with reaction wheels. Until the arrays could be replaced (in 1993), the only solution was to wait until the vibrations died down each time, which dramatically reduced the observing time.

A more critical error was noticed in the images themselves, which displayed clear evidence of spherical aberration – something unimaginable in such a precisely manufactured instrument. As has already been mentioned, mirrors made for ground-based telescopes up until then had not needed the precision figuring required for the Hubble mirror, and testing was made in more than one way. So why did the supposedly more precise reflective null-corrector test fail? Well, without going into a lot of unnecessary detail, it came down to the construction of the test tower required to hold the optics over the mirror. All of the test optics had to be precisely located and aligned with respect to one another, which put severe tolerances on the structure itself.

Rumour has it that the engineers noticed some paint had flaked off one of the ends of the mounting posts and this had been corrected (wrongly) by inserting washers as shims. The washers were the wrong thickness, which made the entire structure the wrong size and thus threw out the measurements made. Whatever the cause, the misalignment of the test tower meant that the mirror was ever so carefully polished to the wrong shape, which resulted in an unacceptable amount of spherical aberration. It is important to note that NASA was unfairly held accountable for the imperfect mirror, which was entirely the responsibility of Perkin-Elmer. Even some of the cost over-runs were beyond the control of the space agency as they were a result of delays caused by the *Challenger* explosion. However, the huge blowout in the telescope construction costs were harder to justify and became even more of an embarrassment once the telescope did not perform as anticipated.

Two separate methods were used to diagnose the error in the HST mirror after it was first discovered. First the images were examined, and optical physicists created a theoretical model to predict the error in the mirror. Second, the null-corrector test stand was examined to find out just what sort of error the misaligned optics would produce. The two calculations agreed – the hyperbolic primary mirror was figured with too deep a bowl shape, with the difference being a mere 2.2 microns at the edge.[20] It doesn't seem like much but in the scheme of things, with the error supposed to be around 0.01 microns, this is huge. Once the problem was identified, however, the question then became one of how to fix it, and for this the optical engineers came up with a clever solution.

The idea was to add two small (~15-mm diameter) mirrors between the secondary and the image plane which were custom made to exactly cancel out the existing error. Just as donning a pair of glasses of the right type can correct for short- or long-sightedness, these corrective optics were made to correct for an error in the mirror shape. During the first on-orbit service mission in 1993, these optics were installed. The resulting change in image quality is shown in Figure 9.4. The correction does not return the telescope to the same level of performance that it was designed to have – no matter what some press releases reported. This should be obvious, since otherwise the telescope would have been designed this way in the first place. However, the performance is close to perfect and certainly good enough that the loss in performance would barely affect the main aims of the telescope.

Figure 9.4: An impressive comparison of images of stars in 30 Doradus. From left to right we have a ground-based image in good seeing and the Hubble Space Telescope images before (WFPC) and after (WFPC-2) corrective optics were installed. Courtesy: STSCI/AURA.

So what about the measurements of the universe – the reason for building this telescope in the first place? The telescope, as originally deployed, proved too aberrated for these measurements, but once repaired it was up to the task. Two teams used Hubble for making measurements and sure enough they got different numbers for the Hubble Constant. Starting from opposing values of 50 and 100 respectively, they slowly began to approach each other until they both settled on something close to the middle at around 70–75 km/s/Mpc (while each team still managed to claim the other was completely wrong).[21] But the measurements went further than this. Conclusive evidence was garnered that showed that the geometry of the universe is flat.

The universe has been expanding since the Big Bang, and this expansion will change depending on how much matter there is. In a simple model, if the mass density exceeds some critical amount, gravitation can slow the expansion to a halt and the universe will collapse at some time in the future. In an open universe, the expansion continues to increase, while in a flat universe expansion always increases, but at a decreasing rate. So our universe is going to expand forever: slowing, but never stopping. This may seem to be a particularly bright future since we are avoiding the annihilation of a closed universe, but the truth is exactly the opposite. The distance between galaxies and therefore stars will keep increasing, with less matter in any given place to form planets and stars. Meanwhile, the suns themselves will eventually burn out into cold cinders. The future is going to be very dark and very cold.

Even when this model of the universe had been established, the Hubble Telescope gathered other evidence that showed that even this slowing of the universal expansion was not proceeding as simply as was first assumed. It turns out that the expansion is slowing down over time due to something which acts like anti-gravity. In fact, this effect was something that Einstein had postulated when he first published his General Theory of Relativity. His original equations suggested an expanding or contracting universe, something which grated against his sensibilities, as he thought the universe should be perpetually static. In an extraordinary step, he added a fudge-factor to his equations known as the Cosmological Constant which acts to perfectly oppose any expansion or contraction. There was no physical basis for adding this constant, and later events showed both theoretically and experimentally that it was superfluous. It just goes to show that even the great scientists get it wrong for reasons of preconceived notions about how the universe 'should be'. Now it appears the cosmological constant is back, though with a much smaller value that Einstein wanted, so it does not completely counteract the universal expansion we see from the flat universe.

From a scientific standpoint, the Hubble Space Telescope has been an outstanding success. It has been possible to use it to complete all of the expected science goals, and the telescope has produced countless unexpected discoveries. Even some images which produced little information of scientific value were sufficiently spectacular that the general public found them truly memorable. In fact there is now a 'Hubble Heritage' website completely devoted to showcasing the wonderful photos taken by the instrument. The many calendars, screen savers and coffee-table books devoted to these astronomical happy snaps are testament to the pure awe-inspiring beauty that HST has brought into the homes of the public at large. The scientific successes still continue as of the publication of this book, but it appears the end may be nigh for this telescope.

As it stands, the telescope is quite capable of operating for several more years at its current level of scientific performance. However, problems with the gyroscopes may occur at any time and prevent the telescope from being pointed accurately. Tests have been made which show that the telescope can operate with as few as two gyroscopes, but eventually repairs will be required. A fourth and final service mission will take place in late 2007 in which the Planetary Camera and Spectrograph will be upgraded and gyroscopes replaced to extend the life of HST to 2010. Making this decision was difficult for NASA as the telescope is not in an orbit which is easily accessible to the Space Shuttle, but given the success of the instrument to date it would be a travesty if this mission did not go ahead. The Hubble Telescope continues to be the source of about 25 percent of all papers published in astronomy, and with better instrumentation, there is every chance that this will continue.

Another reason NASA has given for not attempting an upgrade of HST is that it would prefer to redirect those funds towards the James Webb Space Telescope (JWST). JWST is a $4.5-billion telescope with a 6.5-m diameter segmented mirror; a remarkable telescope on its own merits. However, it is not a replacement or upgraded version of the Hubble Space Telescope. Most important of all, it is an infrared telescope designed to operate from 0.6–28 μm and is *not* an optical or ultraviolet telescope. The difference is significant for several reasons:

1. While JWST has five times the collecting area of HST (and hence

will be able to image much dimmer objects), the resolution will be at best two times *worse* than Hubble's. While resolution isn't everything, none of the images will show any more details than already achieved with HST.

2. HST could (until recently) be periodically serviced in orbit, while the JWST will be parked around a semi-stable point some 1.5 million km from the Earth where it will be inaccessible by humans. There is a good reason for this – JWST should be in a place where the infrared background is a minimum, but it means there is no possibility of repairs or upgrades at any time.
3. It is a minor point, but every image produced by this telescope will be false-colour, so none of them will reflect what they may appear to look like to the human eye. While the Hubble Heritage images are often manipulated and enhanced, they are at least comparable to realistic appearances. JWST will produce impressive photos, but false-colour images rarely capture the imagination of the public in the same way as visible ones.

As of writing this, the progression of events for the construction of JWST is eerily familiar to those for the HST. When conceived in the mid-1990s, it had the name of the Next Generation Space Telescope, which itself promised something that is no longer to be considered. In 2002 NASA renamed the spacecraft after a former director, so perhaps this will go some way to preventing this belief from being perpetuated.[22] Cost over-runs (it had an original price-tag aimed at $1 billion) and constant reductions in size (6.5 m down from 8 m) and spectral coverage have occurred over the last ten years and are certain to continue. Originally it was supposed to be in orbit by 2007, but technical difficulties (some preventable, others not) continue to delay progress. It now has a launch date of 2013, which some insiders still insist is 'ambitious'.

Meanwhile, the future of space-based visible/ultraviolet astronomy is fairly bleak. There are talks of a successor visible-ultraviolet telescope to the HST, but none of the designs is very mature, and certainly no realistic launch date has been set. The most optimistic timeline is around 2020. The only good news is that while a general imaging system may not be on the horizon, new technologies applied to ground-based telescopes have appeared since the launch of HST that may reduce some of the need for a space telescope.

Chapter 10

Advanced telescope techniques

Lightweighting

In Chapter 5 we saw how the design and construction of the telescope had to be precisely controlled to carefully protect the fidelity of the wavefront as it passed through the optical system. In many cases, this is a fairly straightforward process, but as telescopes get larger and larger, optical engineers start to come across new problems that have to be solved. To begin with, consider the mass of the mirror itself. As a general rule of thumb in the past, the thickness of a mirror was set at around one sixth its diameter. The reason was that, for a solid piece of glass, this was the thickness required to prevent it bending under its own mass. To understand this, think of a piece of paper being held vertically. In this orientation, the paper is nice and flat. Now grasp it on either side and hold it horizontally. In this orientation gravity will cause it to bow in the middle. While not as dramatic for a solid piece of glass, this bending may still occur at a level well beyond the quarter-wave Rayleigh limit, so we must thicken the mirror to make it stiffer.

A 1-m mirror will weigh about 250 kg, which is not so bad, but since the volume increases with the cube of the diameter, we can see that a similar 10-m mirror would weigh 250 tonnes. This is about the same mass as the Airbus A380 aircraft. Obviously, supporting and steering such a heavy mass would be an engineering nightmare. Engineers have developed two main methods for solving this problem:

1. Create a thick compartmentalised mirror with much of the glass missing, but enough internal structure to stop it bending.

2. Create a very thin mirror but develop a support mechanism which can compensate for the bending.

Most thick mirrors currently fabricated are no longer solid pieces of glass. In fact, they will have 'cellular' or 'honeycomb' structural designs. They directly mimic their naturally occurring namesakes in both appearance and in their high degree of stiffness with minimal amounts of construction material. The cells might be enclosed or open, but the important point is that the ribs are configured in such a way that the final mirror can resist bending under its own mass. This process of lightweighting mirrors can only be taken so far, and a point will eventually be reached where any further removal of glass will start to reduce stiffness dramatically. On the other hand, supporting a large thick mirror is relatively easy. If such a mirror is assumed to have enough stiffness to resist any bending, all that is required is a mount capable of holding it to the telescope truss.

Active optics

The second approach mentioned above is simply to create a very thin mirror and to support it in such a way that it doesn't deform under gravity like our piece of paper. The problem now shifts to the design of the support structure itself, and a new technique called 'active optics'. This requires a complex electromechanical system attached at many places on the back of the mirror. The support points are pistons which can be pushed or pulled to change the shape of the mirror by small amounts. For example, imagine our thin mirror is tilted as shown in Figure 10.1. With gravity acting on the mirror, it will want to buckle in the middle just like our piece of paper (though much less). In order to correct for this bending, the active supports are pushed in at the bottom to restore the original shape. Of course it is essential to know exactly what sort of bending to expect, so for this a complex mechanical model is devised by the mirror makers. Using this model, the engineers can predict the magnitude and type of bending which will occur when viewing in any given direction, and computers can be programmed to adjust the active supports automatically for the desired correction.[23]

Active optics has made it possible to create very large primary mirrors which have extremely small mass. A primary can now be made quite thin and adjusted in shape while in the mount. It also means that mirrors can be made from less material, which helps with the issue of thermal equilibrium. Generally, we want all the elements of the telescope to be

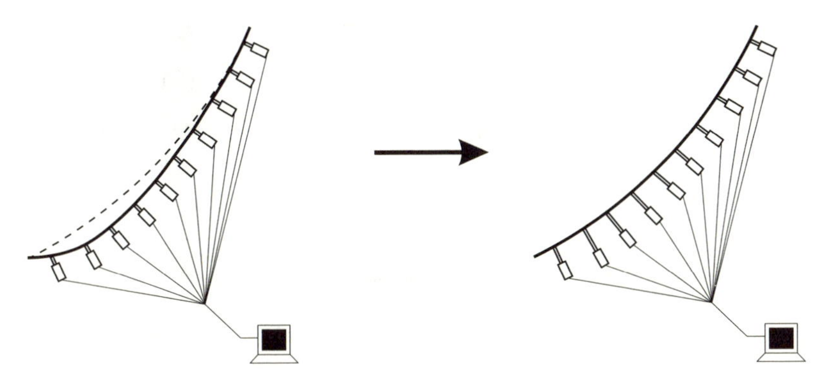

Figure 10.1: The mirror on the left is shown sagging under gravity away from the desired shape (dashed). On the right, the mirror has been returned to the ideal figure by the active control of pistons on the rear.

at the same temperature as the surrounding air so that there is no heat differential in the observatory causing localised air turbulence ('dome' seeing) or condensation. A thin mirror can be cooled down to ambient air temperature much quicker than a thicker one, which gives astronomers more time each night in which to obtain high-quality images.

It may seem complex, but developing a model for controlling the supports on the back is fairly straightforward. Also, you mustn't get the idea that these pistons are moving in and out by large amounts – the graphic in Figure 10.1 is greatly exaggerated. Typically the total amount of corrective motion is a small fraction of a millimetre. Active optics is only required to adjust for deformations formed by varying gravitational effects, and since the telescope changes its orientation very slowly as it tracks objects across the sky, the mirror's actuators are re-adjusted only every few seconds or so.

Other types of active supports can be developed to compensate for bending in the truss, which acts to move the secondary relative to the primary. When pointing vertically, the two mirrors may be perfectly on-axis, but when tilted horizontally, the truss may bend ever so slightly to produce an overall shift and rotation of each element. Active control of the mirrors can be used to correct these motions and shift the optics back to their design positions. Today, all telescopes are built with some form of active support for either mirror shaping or positioning.

Segmented primaries

Active optics has made it possible to increase mirror diameters well beyond what could have been envisioned 30 years ago. However, in

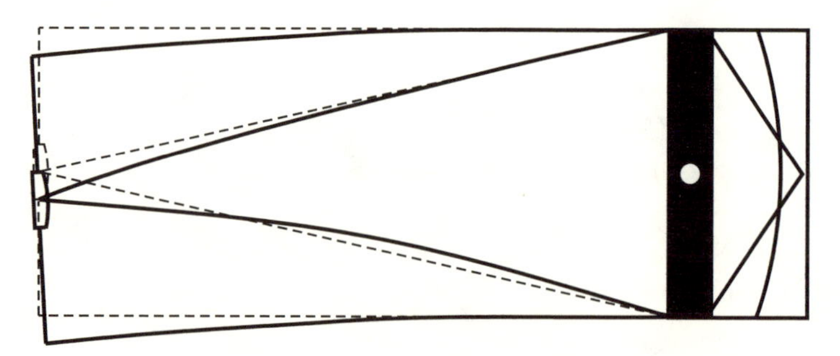

Figure 10.2: Truss bending. A long structure will bend under gravity (shown in solid) which will shift the secondary with respect to the primary and the correct position (dashed).

order to create a monolithic mirror (that is to say, one made from a single piece of glass) there is still the problem of making an oven big enough. Telescope mirrors are made by melting many blocks of glass together to form a single piece, in what is essentially a really big, really hot pizza oven. This is a fairly straightforward process, but it becomes expensive when you have to make your oven 8 m in diameter. Also, there are the problems of making sure the glass melts without any bubbles or cracks; grinding and polishing the huge piece of glass; testing the final surface; transporting it to the observatory – and then constructing a coating chamber. As you can see, the number of big components required for fabrication and operation start to add up. Worse still, each of these has to be custom-made, so the costs quickly balloon out of control and at some point it simply becomes too expensive to consider making a monolithic mirror. The solution is to go to a segmented design.

A segmented mirror is one which is made up of many smaller mirrors which all fit together to act as a single focusing element. One advantage of this approach is that the smaller mirrors can be cheaply fabricated in a small oven, coated in a small chamber and fitted together at a later date. It also means they can be individually removed for cleaning and recoating in a quick and inexpensive manner. Replacement of damaged or underperforming sections can also be accomplished easily. By comparison, if a large monolithic mirror sustains any damage the entire piece of glass must be repolished or even replaced: a very expensive proposition. A further saving can be made in the thickness (and hence mass) of the mirrors. In terms of support, it is a relatively simple proposition to fabricate a 1 m segment which is only 10 mm thick. However, making an 8-m diameter mirror this thick is extremely difficult as it warps in the oven and bends during the grinding and polishing process.

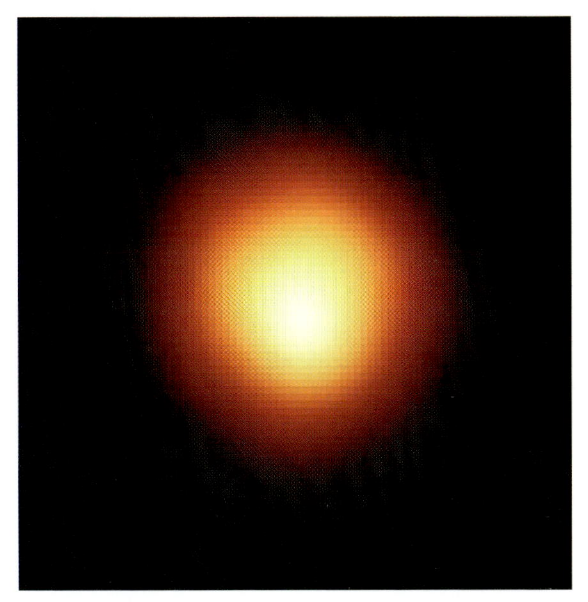

Plate 1: The first image resolving the disk of a distant star. This image is of Betelgeuse, a red giant in constellation Orion. Courtesy: Hubble Space Telescope (STScI/AURA).

Plate 2: An image of the Rho Ophiuchi nebula.
Courtesy: Anglo-Australian Observatory/Royal Observatory of Edinburgh.

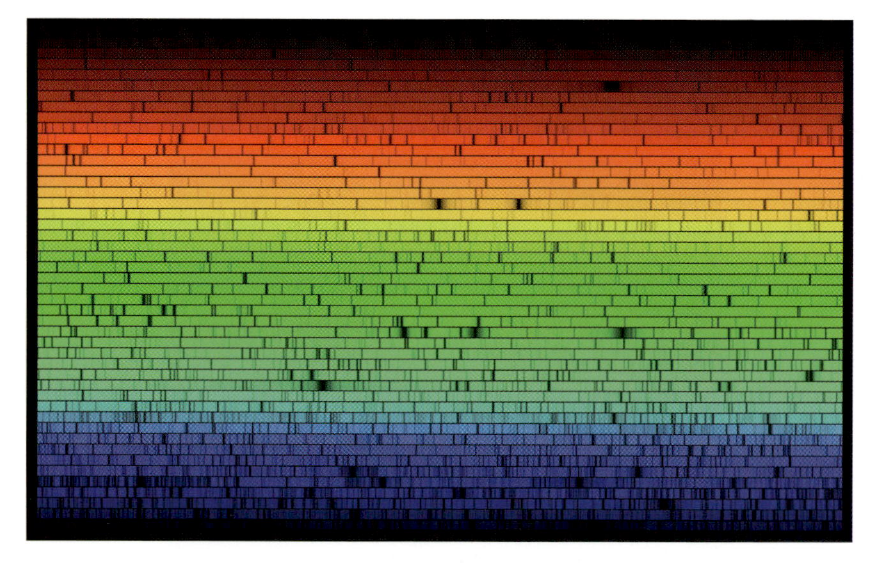

Plate 3: The solar spectrum showing a multitude of dark absorption lines. The entire spectrum can be constructed by joining the right side of a row with the left side of the row immediately beneath it. Courtesy: Nigel Sharp, NAOA.

Plate 4: A guide star laser is shown projected from the dome of Gemini North. The 30-cm diameter laser is about 12 Watts and appears slightly curved only because of distortion in the camera lens that took this photograph. Courtesy: Gemini Observatory/AURA.

Plate 5: A contour image generated by airborne lidar, rendered 26 days after the World Trade Center disaster. The resolution in this image is around 30 cm. Courtesy: NOAA/U.S. Army JPSD.

Plate 6: The Large Aperture Mirror Project (LAMP) is a seven-segment, 4-m beam expander built for the Space-Based Laser experiment.

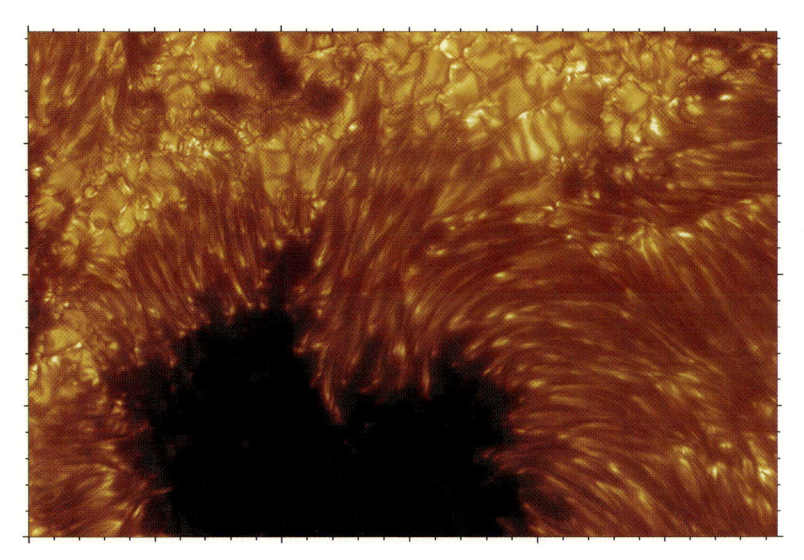

Plate 7: An image of the Sun's photosphere taken with the 1-meter Swedish Solar Telescope. The tick marks around the edge are 1000 km apart, so the Earth is about the same diameter as the sunspot. Courtesy: Swedish Academy of Sciences.

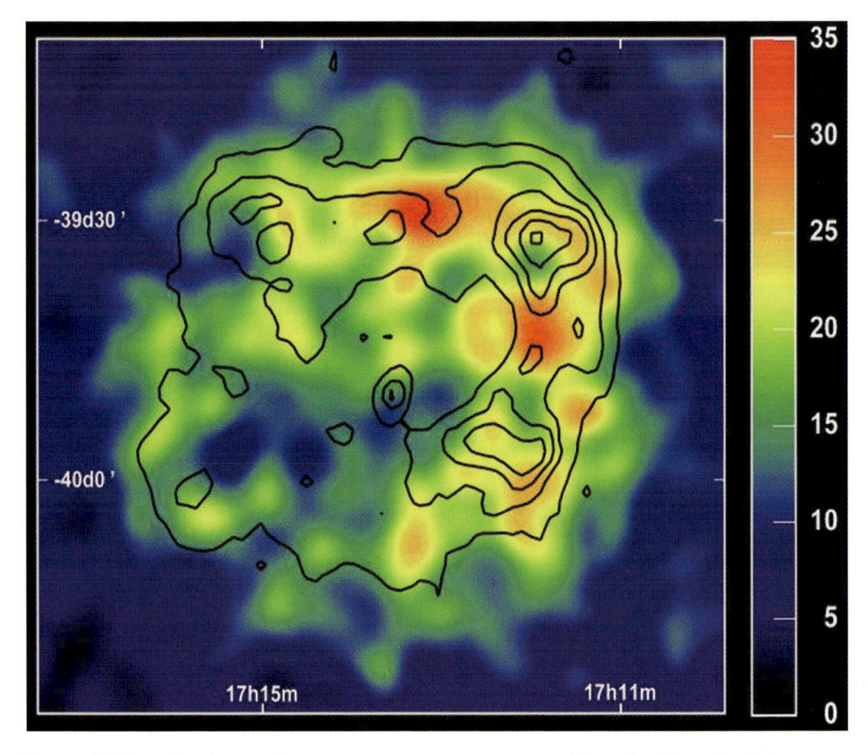

Plate 8: A false-colour image of a supernova remnant as generated from the Cherenkov light resulting from gamma-rays striking the Earth's atmosphere. Courtesy: HESS project.

Plate 9: Pluto at high resolution. Through successive transits of Charon, fine changes in brightness were interpreted to give the above image. Courtesy: Eliot Young.

Plate 10: The first image of an exo-solar planet. The dim planet (red, left) orbits a brown dwarf (white) the distance indicated. Courtesy: ESO/VLT.

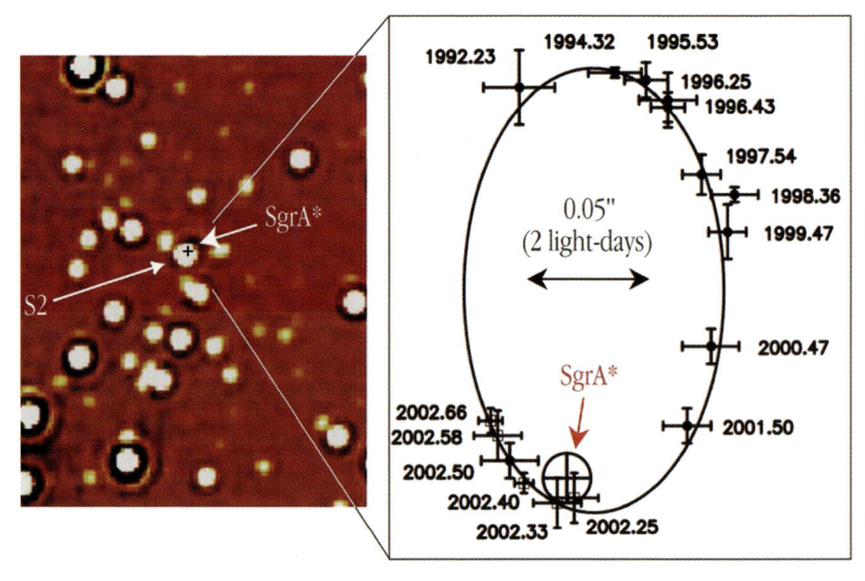

Plate 11: The galactic centre. The image on the left is the star field close to Sagittarius A*, the suspected centre of the Milky Way (shown with a +). Notice the appearance of the brighter stars as nice Airy spots. On the right is a plot of the orbit of star S2 about this point (each measurement represented by vertical and horizontal error bars). Courtesy: ESO/VLT.

Plate 12: Hubble Ultra Deep Field. Every object shown in this image is a separate galaxy. Courtesy: Hubble Space Telescope (STScI/AURA).

Plate 13: Hoag's Object. Courtesy: Hubble Space Telescope (STScI/AURA).

Plate 14: The Pleiades star cluster. The stars have bright four-pointed plus-signs through them as well as haloes which are the result of a spider and a corrector plate in front of the primary mirror. The 'fuzz' around the star cluster is real, and is a beautiful example of a reflection nebula. Courtesy: Anglo-Australian Observatory.

Plate 15: An artist's conception of the Overwhelmingly Large Telescope. Note the roll-away enclosure. Courtesy: ESO.

About now you should be thinking: there must be a catch, otherwise telescope mirrors would have always been made like this. And right you are. The problems come in making the segments and in maintaining a perfect surface. To begin with, there is the problem of making a segment itself. When a single mirror is broken up into segments, most of them will have a different shape. This is because the telescope primaries tend to be either parabolic or hyperbolic and these surfaces have the property that the curvature changes as you move from the centre to the edge. With a parabolic mirror, the segments nearer the centre will have more curvature than the outer ones. Fabricating and testing each individual segment is a long and tedious process compared with figuring a single mirror which can be tested in its entirety in a single step.

It now remains to put the segments together to form a single mirror. In the case of a monolithic mirror, all that is required to ensure it maintains its polished shape is to provide the proper support. With a segmented mirror, the individual segments must each be supported in such a way that they have a perfect surface figure; but added to this, they all have to be aligned to one another in order to produce an uninterrupted smooth shape. Sensing their orientation and position and then correcting for any errors is a difficult process. Furthermore, once they have been correctly positioned, a method must be found to ensure that they stay that way. The 10-m diameter Keck telescopes on Mauna Kea, Hawaii are each composed of 36 segments. Once a week, these mirrors are realigned ('phased') to form a single perfect primary. By observing a star for several minutes, the operators can ensure that each mirror is moved into the correct position. After this has been accomplished, computer control is used to monitor and adjust the relative positions of each segment throughout the night to maintain the perfect focusing.

With advances in telescope design and fabrication, there has been a sudden renaissance in telescope construction unlike anything in the instrument's 400-year history. Until 1990, the three largest telescopes in the world were in the 4-m to 6-m range. The following 15 years have seen the appearance of no less than 14 telescopes larger than this, with most of them in the 8-m to 10-m range. The increases in light gathering were further enhanced by improvements in site selection. As good as they are, however, the ultimate resolution of these telescopes is still no better than you'd get for a cheap commercial telescope. At least, that was until recently when a new technique came along which changed all that.

Adaptive optics

Since the invention of the telescope, astronomers have struggled with the issue of turbulence. Optical construction techniques have advanced to such an extent that even at the world's largest observatories at the best sites, the atmosphere is easily the most limiting factor for image quality. Putting a telescope at as high an altitude as possible will decrease the effect somewhat. The amount of distortion that the atmosphere imposes on a plane wavefront can be modelled, but since the final wavefront surface is random and constantly changing, the precise shape cannot be predicted. We can, however, perform statistical analyses of the turbulence to get some information about what to expect over time. In 1965 Fried defined a parameter, r_o (now called the Fried parameter), which is basically the largest diameter over which an average wavefront can be considered diffraction limited. A sample of a transmitted wavefront any larger than this will have portions exceeding the $\lambda/4$ criterion (as shown in Figure 5.12). The Fried parameter is usually around 5 cm or so for typical city viewing at visible wavelengths, but is more like 15–20 cm at excellent sites such as the peaks of Mauna Kea, Hawaii or Paranal, Chile and possibly as large as 30–50 cm in Antarctica.

The value of r_o simply represents the largest diameter a telescope can be before the effects of the atmosphere start to limit the image quality. To put it another way, a 1-m diameter telescope in the city (where r_o=0.05 m) will have about the same resolution as a 0.1-m diameter telescope at the same location. The reason for building larger telescopes, then, is not to get better resolution, but simply to gather more light so that the objects appear brighter. It also helps us get more photons for our instruments to analyse. Another reason is that many of the world's largest telescopes tend to operate in the infrared as well as the visible part of the spectrum. It turns out that r_o scales with wavelength according to $\lambda^{6/5}$. So if the seeing is a mere 20 cm at a green wavelength of 0.5 μm, then at an infrared wavelength of 2 μm it is whopping 1 m. This is a significant improvement, but it still means that under normal conditions, the 8-m and 10-m class telescopes take images with lower resolution than the 2.4-m Hubble Space Telescope. At least, that was the case before adaptive optics.

In 1953, Horace Babcock suggested that it might be possible to remove the distortions on a wavefront of light by imposing the inverse aberration on the wavefront just prior to the image plane. The principle would work in much the same way as corrective lenses are used to remove focal defects in the eye. In this case, the 'corrective glasses' would be designed for the precise blurring effect of the atmosphere, permitting a large telescope to operate at the diffraction limit. This sounds easy in principle, but there are several problems to be considered in practice. To begin with, the precise shape of the wavefront has to be characterised somehow. To make matters worse, this has to be done quickly and often. The atmosphere is in constant motion, so there is a significant change in the shape of a wavefront every 5–10 milliseconds or so, which means that the shape of the wavefront has to be calculated in a much shorter time. Lastly, some mechanism is required to apply the inverse wavefront error to the beam in order to restore it to a diffraction-limited plane wave. This also has to be done at the same rate as the sensing. The method for this 'adaptive optics' technique actually suggested by Babcock was ingenious, but not really practical or technologically feasible at the time. For this reason, adaptive optics sat in the 'nice idea' basket for a quarter of a century before the US military began secretly working on it.

The progression of the Cold War into a space race suddenly created a need for high-resolution images of orbital satellites. For each side, it was vitally important to know just what the other was putting into orbit. In order to achieve this goal, the US Air Force began to develop techniques to correct for the blurring of the atmosphere. Throughout the 1970s and 1980s, adaptive optics was developed into a workable system, unbeknown to most astronomers. Towards the end of the 1980s, however, the astronomical community began to devote some time and effort to the idea. In 1991, with the Cold War ending, the Air Force decided to declassify the existence of adaptive optics and publicly presented details on how to make it work. Overnight the field of astronomy took a giant leap forward unlike any since the development of mirrored optics.

So how does it work? The first step requires a means for sensing the wavefront distortions. This can be done using many different techniques, but perhaps the simplest to describe is the Shack-Hartmann Wavefront Sensor (SHWFS for short). To understand the operation of this sensor, consider the process of focusing light with a single lens as shown in Figure 10.3. A plane wave entering the lens on-axis will be focused on-axis, while a plane wave coming in at an angle will shift the focus off to one side. The amount of shift and its direction relative to the optical axis tell us how much tilt there is and which way it is orientated. The SHWFS uses this idea to divide the wavefront up into many pieces (called subapertures) and find the slope at each place.

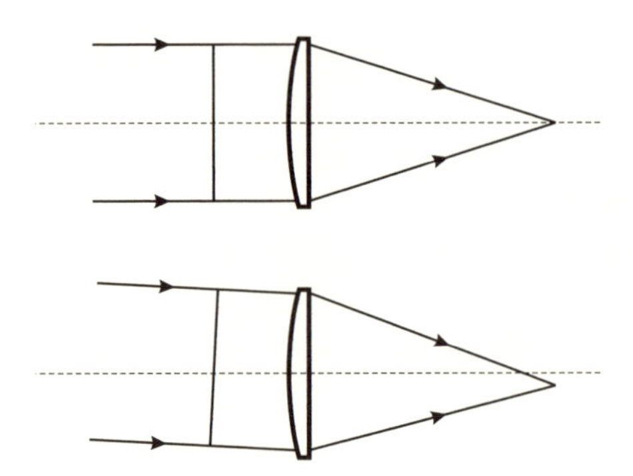

Figure 10.3: Top: A plane wave entering a lens on-axis will be focused to an on-axis location. Bottom: A tilted wavefront entering a lens will be focused to a position off the optical axis. More tilt on the incoming wavefront will cause the focus to shift further from the optical axis.

A SHWFS device consists of many tiny lenses (called lenslets) in an array in front of a CCD. The process begins by observing a point source of light, called a guide star. As this wavefront passes through the atmosphere, it is aberrated into a random shape. When it is directed into our sensor, each lenslet will focus a portion of the light, with the position of the spots indicating the slope of the wavefront over each subaperture (Figure 10.4). The CCD detects where the spots are, and a computer is used to calculate the tilt over each subaperture. By combining the slopes of the wave from each subaperture, we can complete a picture of the shape of the entire wavefront.

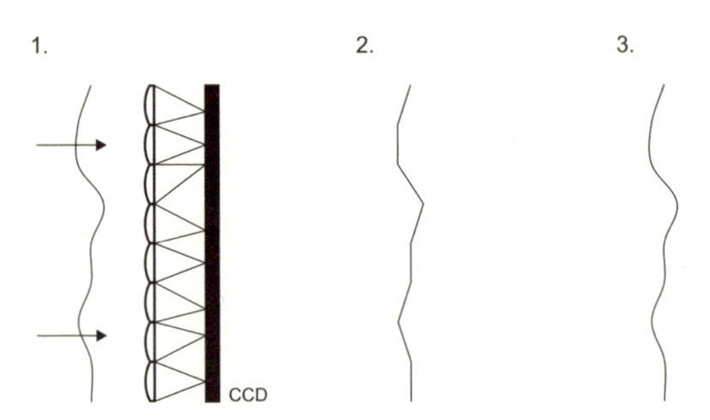

Figure 10.4: 1. When a distorted wavefront enters the Shack-Hartmann Wavefront Sensor, the lenslets create many foci, the locations of which are recorded by a CCD. 2. A computer calculates the slope over each subaperture. 3. Using further algorithms, the overall smoothed shape of the wavefront is found.

Figure 10.4 shows a simplified two-dimensional version of the wavefront sensor, but in reality the analysis is done in three dimensions with an array of lenslets. On the CCD, this might give a pattern of focal spots as shown in Figure 10.5, where you can clearly discern a bowed shape of the wavefront aberration. In practice, the spots will have an irregular arrangement as the shape of the aberration will be quite random. From this we can also see that the choice of the number of subapertures is important. Too few, and the wavefront distortions will not be adequately defined; but if there are too many, we could be wasting computing power being overly precise. Generally speaking, though, there will be one or two subapertures for every seeing-sized disk over the diameter of the primary (i.e. the number will be roughly D^2/r_o^2). Thus a typical 8-m telescope at a good site will require a wavefront sensor with some 400 or so subapertures.[24]

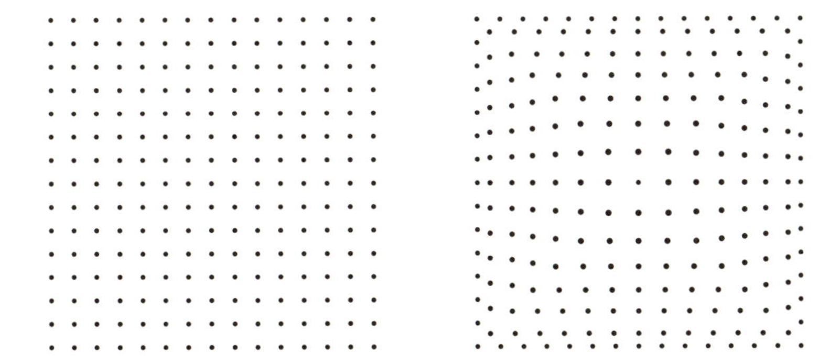

Figure 10.5: A flat wavefront produces a regular alignment of focal spots in a Shack-Hartmann sensor (left). A distortion in the incoming wavefront shows up as a change in the position of the spots (right).

Once the shape of the wavefront is known, we need to remove the aberrations. Again, there a few ways of doing this, but the most commonly used is a deformable mirror (DM). This often consists of a very thin mirror which has many actuators attached to the back, much like a piece of metal foil on a bed of nails (see Figure 10.6). There are different types of actuators, but they are all have the same basic principle: by applying a particular voltage to each actuator, they can be made to push or pull on the mirror. These changes are very small and must be made very quickly – moving as much as a micron in 10 milliseconds. The overall result is that the mirror surface can be deformed to match just about any shape. Likewise, a wavefront of light reflecting off such a mirror can be made to take any shape we desire simply by adjusting the surface of mirror with the actuators. In general, the number of actuators will be about the same as the number of subapertures used in the sensing.

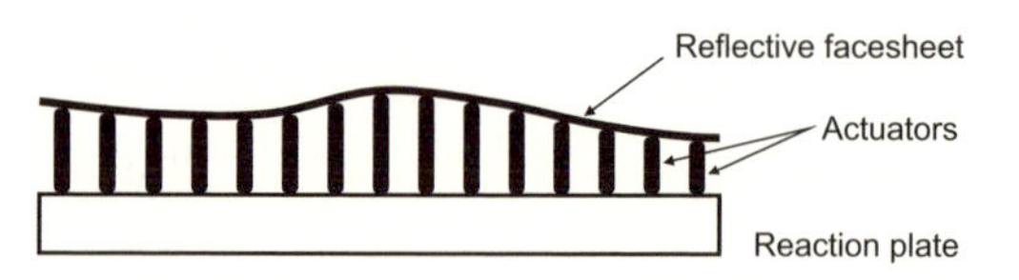

Figure 10.6: A schematic of a deformable mirror is shown, with a top reflective surface formed by adjusting actuators on the rear.

Now we're ready to put this all together to give the entire adaptive optics system as shown in Figure 10.7. Here's how it works: a plane wave of starlight enters the atmosphere and becomes aberrated by air turbulence. We pass the light through a beamsplitter, which reflects a small part of the beam into a wavefront sensor. Once the wavefront distortions have been determined by the wavefront sensor, the inverse distortions are applied to the shape of the deformable mirror.[25] The majority of the starlight which continued straight on through the beamsplitter is then reflected off the deformable mirror to give a diffraction-limited plane wave. This wavefront is finally focused down to an unaberrated image of the star. And all this takes place around 100–200 times every second.

While conceptually simple (and some details have been omitted or altered to help simplify the description), it took many years to produce a workable system. One of the major limitations during the late 1970s was the computing power required to make the necessary calculations and corrections every few milliseconds. Nowadays, of course, this can be handled by a standard personal computer. Even so, most observatories still operate in the infrared rather than the optical as it reduces the number of subapertures. Another problem is that the guide star must be very bright in order to be suitable for adaptive optics. Only a small amount of the collected light is sent into the wavefront sensor, as we wish to have most of it going into the imaging camera. Of this reduced light, only a small portion is intercepted by each subaperture. Furthermore, a wavefront measurement must be made every few milliseconds. Taken together, these limitations mean that we find that in order to have enough light for the wavefront sensor to make a reliable measurement, the guide star must be quite bright. The simplest adaptive optics systems can use a bright natural star for this

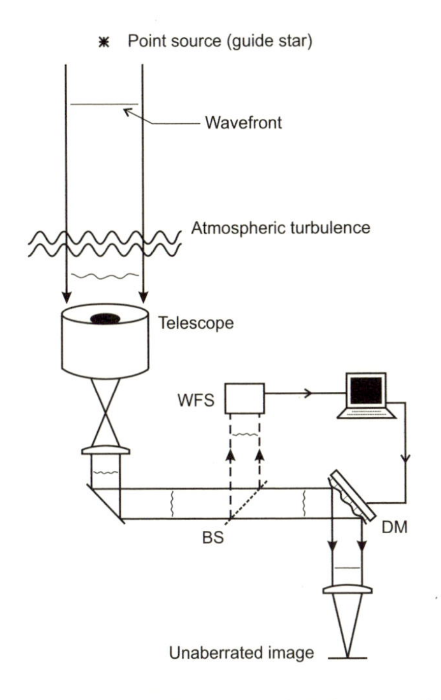

Figure 10.7: An adaptive optics system. Aberrated light from a guide star is collected by a telescope then collimated. The light is directed through a beamsplitter (BS) that picks off some of this light into a wavefront sensor. The aberrations are computed and the deformable mirror (DM) given the correct shape to correct the light into a plane wave for diffraction-limited imaging.

purpose, and one of the first such systems used on an astronomical telescope was attached to the 2.5-m Hooker telescope on Mt Wilson.

I visited the site in the first week of September 1996, only a year after the system was first commissioned. I was being shown the system by Chris Shelton, one of the people instrumental in taking the newly declassified technology into the civilian world. To the casual observer, the system components looked rather like a jumbled mass of wires and mirrors attached to an optical bench. This bench was itself attached to the side of the telescope in what appeared to be a very precarious position. For simplicity, the adaptive optics bench was attached to the side of telescope because of the equatorial mounting. These days, since most telescopes are on alt-az mounts, the optics are located on a separate Nasmyth platform or in a completely different room from the telescope, with the light relayed via several mirrors.

On this particular night we were lucky enough to have cloudless skies and the system was up and running. Grateful to be inside the warm control room, rather than out in the very cold night air of the main dome, I was concentrating on getting some feeling back into my hands while Chris began taking images. Oddly enough, the late hour was no real problem since I was still suffering jet-lag after flying in from Australia the previous day. These distractions left me as I was treated to a couple of tiny dots on the computer screen. Chris was grinning from ear to ear, but I will admit to being a little less than impressed. I mean, binary star systems are a dime a dozen, and if you've seen one you've pretty much seen them all.

Perhaps sensing this, Chris provided a little more information. He explained that the two stars were omicron Ceti, otherwise known as Mira. Long believed to be a single star, it was only identified as a binary using this very telescope in 1926. Even then, it was only possible to infer this by isolating two separate spectra from each of the stars – one a red giant, the other a white dwarf. What we were seeing was what no one else had seen before; the very first proof in the form of an image showing the two separate stars. This was only possible using the adaptive optics, permitting us to resolve the 0.6-arcsecond separation between them. It dawned on me that this was what it was like to be an astronomer – seeing something no-one else had seen before.[26] I began to see how you could get addicted to this stuff.

Figure 10.8: A picture of the ADOPT system on the 2.5-m Hooker telescope (left), and Mira clearly resolved as two individual stars (right). Courtesy: Mt Wilson Institute.

Adaptive optics works, but there are limitations. To begin with, there is the issue of field of view. When we analyse the wavefront of light from the star, we are detecting the aberrations it gets passing through a particular column of atmosphere. These are the aberrations we correct with the deformable mirror. Light from an object somewhere off-axis will pass through a different column of atmosphere and thus have a different aberration (Figure 10.9). The deformable mirror will be unable to correct for this different distortion, so there will be some residual aberration component. This error is known as angular anisoplanatism and results in a limited field of view over which we can use the adaptive optics. Images of objects outside the so-called isoplanatic angle will benefit somewhat from the correction applied by the deformable mirror, but to a lesser extent (decreasing as we get further off-axis). Typically this angle is quite small – around half an arcminute or less for a large telescope.

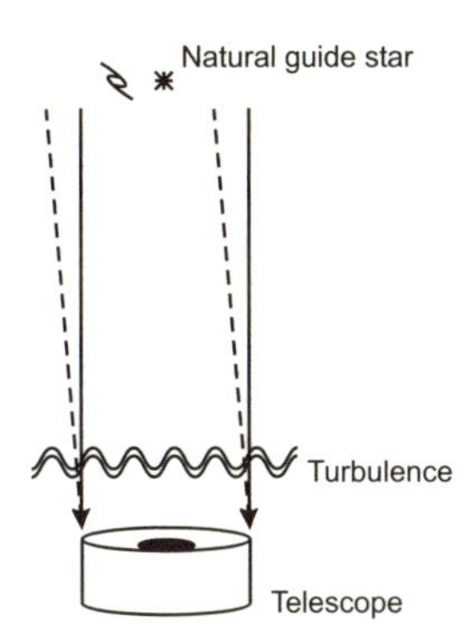

Figure 10.9: Angular anisoplanatism arises because light from the off-axis astronomical object (shown dashed from a spiral galaxy) travels through a slightly different portion of the atmosphere to the light from the on-axis guide star.

The isoplanatic angle puts limits on how far the guide star can be from the object we wish to study. Meanwhile, the wavefront sensor sets limits on how dim the guide star can be. With these two factors combined, there are only a few natural guide stars which are suitable for adaptive optics and it is rare to find one sitting close enough to an object we really wish to observe. If this were the entire story, adaptive optics would probably be relegated to the realm of an exotic technique with limited use. Fortunately, however, we have found a way to generate artificial guide stars using high-power lasers.

Laser guide stars

At around 90 km above the Earth's surface, there is a tenuous layer of sodium atoms. Sodium has a particularly strong absorption line (actually a doublet, or two closely spaced lines) at a yellow-orange wavelength of 589 nm. If we direct a bright laser tuned to this wavelength into the sky, we can excite the sodium atoms and cause them to glow. In doing this we can generate an artificial laser guide star (LGS). This solves the problem of limited sky coverage from natural guide stars, as the laser guide star can be produced in any direction we point our laser (and telescope). A photograph of a guide star laser operating at the Gemini North telescope in Hawaii is shown in Plate 4.

The technique is not without its limitations. The major problem comes from the fact that the artificial star is relatively close to the telescope. In general, the bulk of atmospheric turbulence occurs at an altitude of around 8–10 km. While the laser guide star is formed higher than this, the light from the objects we are interested in viewing comes from infinity. Thus the light from the laser guide star will pass through a slightly different atmosphere than light from objects at infinity. This can be seen in Figure 10.10 where the light from the laser guide star forms a cone through the turbulent layer. The effect is called focus anisoplanatism (or cone effect), and puts limits on how much correction the laser guide star adaptive optics system will provide. The amount of correction we can get also decreases as telescopes get larger.

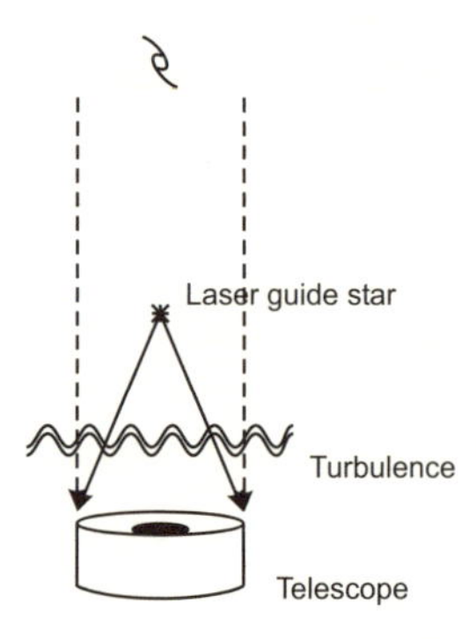

Figure 10.10: Focus anisoplanatism arises when using a proximal point source for wavefront sensing. Light from the laser guide star travels through a different column of atmosphere than does the light from the astronomical object (dashed).

A typical laser guide star requires at least 10 Watts of laser power at a wavelength of 589 nm. Unfortunately, generating this laser wavelength is not a simple matter, and the high-power requirement makes it even more of an engineering challenge. Only recently has it become possible to produce such a laser, and still the cost is around a million dollars or so. At a typical large observatory, the laser is located in a room separate from the main dome. This is done so that the very warm laser will not heat the air inside the dome (and so cause dome seeing) and we can eliminate the possibility of stray light disturbing sensitive observations. The light is then piped into the dome (either by mirrors or inside an optical fibre) and directed into the sky. Systems like this are coming online (with minor variations) at major observatories today in order to provide near diffraction-limited quality imaging, and we will see examples of images they produce in later chapters.

Of course, where one guide star is a good thing, more are better. With this in mind, telescope designers are beginning to contemplate a new generation of adaptive optics using multiple guide stars. The increased complexity and cost have several benefits apart from the improved job security for the optical engineers. The general idea is that you effectively probe different columns of air above your telescope to completely characterise what the entire atmosphere is up to. Multi-conjugate adaptive optics, as it is known, should make it possible to eliminate most of the cone effect and let astronomers have a near crystal-clear image over a wide field of view. The calculations are a lot more complex, and of course there is the need for as many as five 10-Watt lasers. Given that you can see these orange beams from a few kilometres away, the appearance and performance of the observatories in the future is sure to be very impressive.

Chapter 11

Laser communications and remote sensing

Telescopes are instruments for gathering and intensifying light. Until now, we have primarily considered using them for astronomy, and you would be forgiven for thinking that this is their only use. In fact, telescopes are used in many other areas of research, such as remote sensing and laser communications. Remote sensing is simply a general term we give to the analysis of light from distant sources and includes laser radar as well as surveillance. In this chapter we will look at some of these applications and how they benefit from improved telescope technology.

Laser communications

Laser transmission involves the projection or gathering of laser light from some source. A laser is simply a device that can generate light of a single wavelength (monochromatic) where all the waves are in step with one another (coherent). This gives the beam the unique property that it will not spread out quickly. Such a small divergence makes lasers ideal for transmitting signals efficiently since all of the power is directed towards the receiver. Our first application thus begins with optical communications. To understand why we would want to use light to transmit data signals, we have to understand how a wave can carry information. Take a look at the recording of two sound waves in Figure 11.1 and notice how the lower one has a lot more variation and detail to it. As with any picture or book, the more detail we see, the more information is being given. In this case the first waveform is from a single instrument while the second is from an entire band.

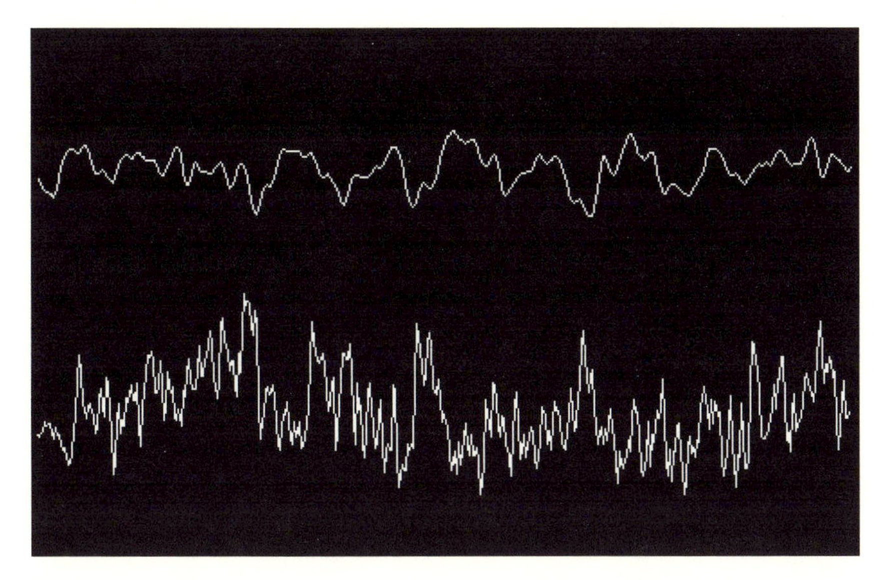

Figure 11.1: The lower waveform contains more information than the upper one, as can be seen by the increased detail.

The detail comes in the form of more rapid changes in amplitude over smaller timescales. This increased waveform complexity (called bandwidth), requires the use of shorter and shorter wavelengths for transmission. Radio waves were the first method developed for electromagnetic transmission of information, as the signals spread out in all directions and over vast distances. This is ideal for radio stations where a simple audio signal with low informational content (such as talk-back radio) is to be distributed over a large area. If we wish to transmit larger amounts of data, such as thousands of simultaneous phone calls or real-time video over the internet, radio waves cannot provide sufficient bandwidth. For this, we need to go to much shorter wavelengths such as infrared or visible light. This is why telecom companies are busy putting optical fibre into the streets: piping optical signals from one place to another provides higher bandwidth information.[27] Optical fibres may be ideal for transmitting signals between houses, cities and even countries (as in the case of cables laid out on the ocean floor). But they are not practical when you want to communicate with a moving source such as a plane in flight or a satellite in orbit. Communication of this sort has been traditionally accomplished using radio waves or microwaves, but these days we wish to send larger amounts of data. As such, many governmental and commercial groups have been looking into the concept of transmitting laser signals over vast distances of open air.

The concept is not new. Since the advent of the laser, there has been a lot of research into so-called 'free space' optical communications. The problem is that the advantages of using laser beams can also be disadvantages. For instance, the low beam divergence means there is no wasted energy in sending signals to the wrong place. It also has the benefit of increased transmission security, since you have to be in the line of sight to intercept the signals. However, the same low divergence puts severe constraints on the pointing accuracy of the transmitter and receiver – if your aim is just a little off, your laser beam will miss being picked up by the receiver. Also, whereas radio signals aren't dramatically affected by the atmosphere, atmospheric turbulence can distort a beam of light to such an extent that it is difficult to focus properly. Advances in laser diodes, precision tracking systems and adaptive optics have begun to make a difference in this field to such an extent that we can now seriously consider transmitting lasers to satellites. In fact this is just what NASA had in mind for a future Mars mission.

Interstellar probes cost millions or even billions of dollars to launch. Each generation of probes carries ever more sophisticated instruments designed to provide information about the object under study. With rapid improvements in sensor technology, we now have the ability to launch planetary probes with a large number of gizmos to record incredible amounts of data. The problem is how to get this back to Earth. Take, for example, the Cassini probe currently in orbit around Saturn. It has five cameras, an imaging radar, a magnetometer, a plasma spectrometer, a mass spectrometer, a dust sample analyser and a magnetospheric imager. As if all that wasn't enough, it also dropped a probe onto Titan which carried its own instruments. All of the collected data gets sent back through a huge antenna transmitting at just 0.9 kilobits/sec – some 50 times slower than a dial-up computer modem – and we all know how frustrating it can be to use these download rates in order to view a simple webpage.

For Cassini, the solution to this bottleneck is to use advanced data compression schemes to reduce the amount of data which is sent. Often data is simply not recorded in the first place since it could never be transmitted. The signals sent back by a probe such as this are extremely weak and require a network of very large radio receivers dotted around the world. This becomes a very costly exercise for long-term missions, and can result in the termination of a mission even while the probe is still active and capable of sending back data. NASA is currently evaluating the cost benefits from continued operation of the Voyager probes for just this reason. The most distant man-made objects continue to send back small amounts of data on the nature of the Solar System's edge (taking around half a day to reach us). But someday the decision will have to be made that it simply isn't worth the $4 million a year required to 'talk' to these objects. It seems a shame, but then they will eventually fall silent anyway when they run out of power sometime around 2020.

The Mars Odyssey probe, currently in orbit around the world of wars, is capable of transmitting data at 128 kb/sec (much greater than Cassini due to some improvements in technology, but mostly because of the much smaller distance). Until recently, NASA had plans for a Mars mission in 2009 which would use laser communications to send back data at a rate of 10 Mb/sec. Unfortunately the Mars Laser Communication Demonstrator (MLCD) was cancelled in mid-2005, but we can still look at some of the technological issues involved. The idea was to have a 10-Watt green laser projected to Earth using a telescope with a diameter of 0.3 m. Using the resolution formula, we can estimate the average diameter of the laser beam at the Earth to be around 900 km. If you can imagine all the light from a dim 10-Watt light globe distributed over an area the size of France, you may begin to appreciate the difficulties involved in detecting this signal.

Because of the weak signal, you need a large telescope on the ground to gather as much of the light as possible. In the case of MLCD it was to be the 5-m Hale telescope at Mt Palomar in California. Additionally, Mars is not always visible in the night sky, so it may be necessary to look for the signal during the day. Worse still, there may be times when the red planet is very close to the Sun. Pointing a telescope at the Sun is never a good idea, and astronomers get very nervous at the thought of even opening their dome in the daytime. When looking near the Sun there is always the danger that the sunlight will reflect off the primary mirror to a focus off-axis. This could lead to a serious risk of damaging equipment in the observatory or sensitive detectors inside the telescope.

Lastly, we have to consider the pointing requirements on the space probe – it has to point the telescope to a receiver to an incredible angular precision. You may think that with a spot nearly 1000 km in diameter, this should be easy to do, but from the distance of Mars, it is the angular equivalent of picking out the width of a human hair at arm's length. But remember that this is an all-or-nothing deal – you must hit your target or you won't get a signal. All in all, optical communications over these sorts of distances are a tremendous challenge and explains why it hasn't been tried yet. Unfortunately, it appears that it will be more and more necessary as we contemplate ever more complex instrumentation on probes. We can only hope that this idea will get funding in the future.

Lidar

Light can be used for communications, but can also be used as a method for detecting certain qualities about a distant region – something described under the umbrella term of remote sensing. Lidar (an acronym for **Li**ght **D**etection **A**nd **R**anging) is just like radar except that it uses pulses of light instead of radio waves. Sometimes called laser radar, laser ranging or ladar, this technique is used in many applications. The principle is quite simple: a pulse of light is emitted by an observer, some of which will reflect or scatter off a distant object. The return signal is collected by a receiver and analysed for the specific quantity we are trying to detect. We can work out the distance (range) to the object as the speed of light divided by twice the time the pulse took to return.[28]

As a simple first example, consider lunar ranging. In 1969, the crew of *Apollo 11* left an array of retroreflectors on the lunar surface. This device consists of many corner cubes (Figure 11.2) which will reflect incident light back to the source over a wide range of angles. It sounds a bit like a mirror, except that when you look into one of them you will always see a reflection of the pupil of your eye, no matter how you twist and tilt it. By sending out a laser pulse from the Earth we can detect the weak reflected light and measure the distance to the Moon. Research has been carried out for many years now using this retroreflector and others which have since been placed at various locations on the lunar surface.

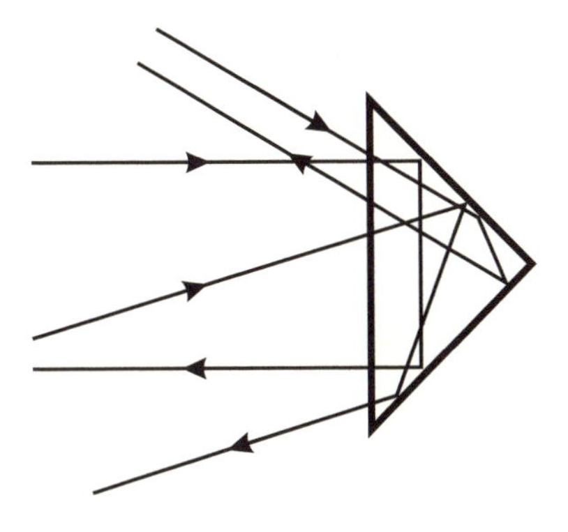

Figure 11.2: A single corner cube retroreflector (left) returns incident rays back to their source. An array of retroreflectors was placed on the lunar surface by *Apollo 11* astronauts (right, Courtesy NASA).

The McDonald Laser Ranging Station (MLRS) experiment run by the McDonald Observatory in West Texas is one such example, in which high energy pulses of green laser light are directed to the Moon. The pulses return 2.5 seconds later, and are collected by a 0.75-m telescope. Even with such a bright pulse (a single 1.5-Joule laser shot can burn human skin) the return signal amounts to only a few photons per shot. From the measurements made over the past three decades, we now know that the Moon is slowly getting further away from the Earth at a rate of 3.8 cm per year. The reason for this is that the Moon produces tides on the Earth that experience a small amount of drag (friction) as they move over the Earth. This in turn reduces the angular momentum of the Earth-Moon system and the Moon moves further away from the Earth (since objects in a more distant orbit move more slowly).

While the rotation of the Moon is 'locked' so that it always shows the same face towards the Earth, the lunar ranging data has shown that the Moon has many different subtle wobbles. One such wobble mode is called libration and allows a ground observer to see up to 59% of the surface of the Moon over time. By studying the various wobbles the Moon has, astronomers have deduced that it must have a liquid core. They have also precisely determined the lunar motion over many years. The constancy of this motion indicates that the force of gravity is extremely stable – neither getting stronger or weaker over long periods of time. This sort of information is very useful to many astronomers who look at the most distant objects in order to model the way the universe behaves. New experiments being carried out by lunar ranging facilities are aiming to improve the accuracy of measurements and test certain aspects of the General Theory of Relativity.

The same ranging technique can be used to find the distances to satellites in orbit around the Earth. Some examples include the Laser Geodynamic Satellites (LAGEOS) and Starshine (Figure 11.3). The data collected in these studies is used to better understand the Earth-space environment and its effects on orbital objects and even monitor the motion of plate tectonics. For example, the Earth's atmospheric density decreases rapidly with altitude and in a low Earth orbit (100–200 km up) it is miniscule, but still not quite zero. Satellites orbiting through this tenuous atmosphere will experience a tiny amount of drag which will eventually pull them back to Earth. As a result, long-term experiments and facilities (such as the International Space Station) must be periodically given a boost to return them to a higher orbital altitude. Without this boosting, satellites and debris simply fall back to Earth – either burning up on re-entry or crashing into the ground. It was precisely this atmospheric drag which caused NASA's Skylab to crash down on Western Australia on 11 July 1979.

Figure 11.3: Examples of satellite ranging experiments, LAGEOS (left) and Starshine (right). Courtesy: NASA.

Laser ranging of the Earth's surface can be made from airborne or orbital platforms. Such instruments can provide us with accurate contour maps of inaccessible or obscured terrain. The principle is the same as for ranging from the ground up where precise timing can give very accurate distances to a distant surface. This principle has been applied to airborne oceanographic profiling of subsurface coastal details. It also proved useful to the United States' National Oceanic and Atmospheric Administration (NOAA) when it flew an aircraft above the World Trade Center site soon after the events of 11 September 2001. The lidar data was used to generate a highly accurate three-dimensional picture of the site, providing authorities with more information to help them carry out relief and clean-up efforts (Plate 5).

Laser ranging of the type described above is very much like conventional radar or sonar in that it is simply used to measure distances in order to build a three-dimensional picture of a remote region. There are other types of lidar, however, in which the emitted light interacts with the target and is altered in a way which can provide us with other information. For example, if we reflect light off a moving object, the reflected signal will be Doppler shifted and change wavelength according to the velocity of the target. This is the principle behind some police laser radars used to measure the speed of cars on the road. If the 'target' is the atmosphere, Doppler lidar can be used to measure wind speeds. In this case, the measurements of distance are made more than once. By dividing the received light up into chunks of time, we can assign each chunk to a different distance (called binning). With binning it is possible to get wind speed measurements over many points through a large column of atmosphere using a single laser pulse.

Using Doppler lidar, it is possible to scan a portion of the sky and map the way winds are moving. The caveat here is that the lidar can only measure the component of wind speed along the line of sight, and is insensitive to cross-winds. But such systems can be used around airports where invisible turbulence and wind sheers can play havoc with aircraft. Since planes have low velocities and altitude at take-off and landing, it is important to know whether strong wind sheers are present which could cause crashes. In the future it may also be possible to install such lidars in aircraft themselves to detect and avoid turbulence a long distance away. This would not only reduce fuel consumption, but would also ensure that no-one in first class spills a drop of their Dom Perignon.

Lidar can also be used to measure temperature. This can be done using many processes, but the simplest and most common involves Rayleigh scattering, where a photon of light scatters off an air molecule without a change in energy (much like the collision of two billiard balls). With Rayleigh lidar we send out a pulse of laser light and simply look for the return light at the same wavelength, and the strength of the return signal gives us information about the temperature of the gas. To see how this works, imagine a box filled with air that is being probed by our laser. The cooler a gas gets, the higher its density and the greater the scattering we should expect for a laser entering the box, as a photon will have more chance of encountering a molecule. The more scattering there is, the stronger the return signal. Thus, we can shine a laser at a distant region of the atmosphere, and know that a stronger return signal implies a lower temperature. Rayleigh lidar can thus be used to remotely sense the temperature of a column of atmosphere.[29] Unfortunately, dust in the lower atmosphere can confuse the measurements, limiting the use of this technique, but it has been useful in improving our knowledge of the upper atmosphere.

These are only some examples of lidars. They can also be used for measuring cloud coverage and altitude; detecting aerosols and other impurities in the atmosphere; for weapons and chemical inspection; for tracking and measuring of plate tectonics and even testing of predictions of relativity. The one common requirement for all lidars is a laser source and a telescope receiver. For atmospheric lidar, where the weak return signal is coming from molecules of gas, the telescope must be as large as possible for even the best detectors to make useful measurements. A larger collector also makes it possible to reduce the power of the laser to levels where they are eye-safe. In the next chapter we will see how liquid mirror telescopes can meet this requirement.

As a last word on the subject, it is worth noting that lidar can be used for surveillance. Sound is transmitted by vibrations in the air which can cause vibrations in surrounding objects. For example, if a conversation is taking place inside a room, the windows will vibrate ever so slightly. By aiming a laser at the windows, the reflected signal can be gathered by a telescope. The phase of the reflected wave can be detected by mixing it with the output laser, with the result being a waveform that represents the vibrations in the window. This technique can be used to detect the sounds and conversations taking place behind closed doors (or windows, as the case may be). So effective is this technique that in places where truly classified discussions take place, the rooms must be well isolated from the outside world. Where this is not possible (e.g. the Oval Office of the White House), there are usually double panes of glass to dampen the signal and active shaker systems applied to the outer pane to further confuse any potential eavesdroppers.[30] As we will see, though, the world of surveillance is not simply limited to lidars.

Chapter 12

Surveillance

Airborne surveillance

From the day the telescope was invented, its implications for the surveillance community have been apparent. The initial application was in providing advance warning of an approaching army or navy, making it possible to prepare a response. Eventually, limits imposed by the terrain or the curvature of the Earth (the horizon) had to be overcome in order to gain any further advantage. Observation towers provided some help, but they could only be located in friendly territory.

With the invention of the hot-air balloon in 1783, it became possible to conduct telescopic observations of vast tracts of land. Both manned and unmanned balloon-borne observations of enemy territory were used with varying degrees of success, but for the most part the concept had limited use. Balloons are at the mercy of prevailing winds and detailed observations could not be reported until the observer returned to earth. Aerial surveillance did not really take off until the invention of photography and the development of the aeroplane. The first practical demonstration came with the photographs taken over the battlefields in Northern Europe during World War I, which clearly showed the locations of enemy trenches (for all the apparent good it did for either side). It is important to note that although the standard terminology introduced at this time called them 'cameras', the telephoto lenses used were essentially refractor telescopes.

In 1936, England initiated a photoreconnaissance project to obtain a complete pictorial record of Germany in anticipation of the intelligence requirements of a future war. As part of this programme, an Australian,

Frederick Cotton, carried out numerous flights over Eastern Europe in the guise of civilian business flights in a private plane. The aircraft was actually fitted out with an advanced camera capable of taking many rolls of photographic images over large portions of the ground below. These were to prove invaluable in later years. When war finally did break out, intelligence gathering kicked into high gear with rapid developments in lens and film technology by both the Allied and the Axis military. Better information on the state of enemy personnel and armaments were critical to organising more efficient offensive and defensive counter-operations.

With both sides appreciating the importance of these reconnaissance missions, increased enemy firepower was directed towards these aircraft to prevent such data gathering. This, in turn, led to improvements in camera and aircraft design to allow them to fly faster and higher. Surveillance pilots even became famous figures in their own right. The French writer/adventurer Antoine de Saint-Exupéry flew many missions over occupied territory for the Allies before going missing in 1944, on what was supposed to be his last flight. Unfortunately, these accomplishments are often overshadowed by a children's book he wrote, which is widely used to take the fun out of learning French.

At the end of World War II, and with the Cold War beginning, development of image surveillance techniques continued unabated. However, with no actual hostilities taking place, the image data collection had to be conducted using more covert measures. In the 1950s, the United States developed the U2 spyplane, which first flew over Soviet airspace in 1956. This high-altitude aeroplane could fly beyond the reach of Soviet surface-to-air missiles (at the time) and had the capability of taking high resolution images with lenses of up to a metre in diameter.

The value of such imagery was never more apparent than during the Cuban Missile Crisis in late 1962. The Soviet Union had installed nuclear missiles in Cuba – some 150 km from United States' soil and a mere 20 minutes' transit time from Washington DC. As proof of these claims, President John F. Kennedy ordered the release of photos taken during a U2 reconnaissance operation. These photos (shown in Figure 12.1), were shocking to most Americans, as most people were completely unaware of either the existence of such a surveillance capacity on the part of their own government, or the amazing detail in which images could be captured. Until this time, airborne (and space-based) surveillance capabilities had been a well-guarded secret.

Even the capture of the U2 pilot, Gary Powers, in 1960 was not widely understood to be part of an extensive US surveillance programme.

Figure 12.1: An image of the Cuban missile launch site taken from a U2 reconnaissance aircraft. Courtesy: JFK Library.

Space-based surveillance

The Cuban Missile Crisis, the capture of Gary Powers and the Soviet Union launch of *Sputnik* all came within in a period of five years. Each of these events played a part in the subsequent reconnaissance space race. Both superpowers soon began massive programmes into space-based surveillance of foreign territory, as space-based telescopes have many advantages over aerial systems. For instance, orbital objects have unfettered access to the region directly above any point on the Earth. This was never more apparent than when *Sputnik* made its first pass over the United States.[31] Another benefit is the access to huge swathes of ground in a small amount of time. While aeroplanes require refuelling and travel at relatively slow speeds, an object in low Earth orbit can completely circumnavigate the globe in less than 90 minutes.

This benefit can also be a drawback; lack of dwell time (or time spent over a particular point on the Earth) and high relative velocity. A space telescope can be moving at a relative velocity to the ground of around 8 km/sec, so images will become severely blurred without a carefully designed tracking/slewing system.

The United States launched its first successful space-based reconnaissance satellite in 1960 on the same day that Gary Powers was found guilty of espionage by the Soviet Union. Part of the Corona programme, the satellite was designated *KH-1* (for *Keyhole*). The telescope consisted of a 0.6-m diameter refractor capable of taking photos of the ground to a resolution of about 7–8 m, recorded on rolls of film a kilometre long. The trick to recovering these photos was an elaborate system designed to eject the film canister for retrieval in mid-air by an Air Force C-119 cargo plane. The Corona programme provided invaluable information for the US and helped disprove the military assessment of a large nuclear weapon disparity between the superpowers (the so-called 'missile gap'). The Corona spacecraft continued to develop, and when the programme ended in 1972, resolution had improved to 2 m with film lengths up to 400 km!

From this point on, most of what we know about space-based surveillance capabilities comes from non-official sources. The National Reconnaissance Office (NRO), a joint USAF/CIA agency which operates spy satellites in the United States does not acknowledge any programmes beyond Corona.[32] However, it is well known that there have been steady improvements in the diameter and sensitivity of the telescopes for greater resolution and spectral coverage. Most important of all, the development of electronic detectors has made fast image acquisition possible, and also provided the ability to use far infrared wavelengths to 'see' through camouflage. Digital media also provided a simple means for the conversion of light directly into electrical signals. Such signals could then be immediately transmitted to the ground. This in turn provided faster image analysis compared with film and longer spacecraft lifetimes since the spacecraft can never run out of film.

So just what can a spy satellite see on the ground? Well, obviously the actual dimensions, specifications and orbital elements of current instruments are highly classified. However, by using the physics of telescopes and some educated guesses, we should be able to get a good idea. To begin with, a telescope will get its best resolution from being in a low Earth orbit (LEO) which is, typically, 100–500 km. Often the observational altitudes can be reduced even further by using highly elliptical orbits, but for the moment we will consider a reasonable value of 200 km. The next issue to consider is the operating wavelength. In order to best 'see' people and engines, most observations would ideally be carried out in the infrared for imaging during the night. But for the best resolution, a shorter wavelength is desirable, so we'll use green light at around 500 nm.[33]

The last unknown is the mirror diameter. Again, we can make an educated guess. First of all, we already know that the 2.4-m diameter Hubble Space Telescope was put into space in 1990 (and built much earlier), so we can be sure the current state of technology is far superior to this. Furthermore, in 1987 the US Air Force constructed a space-ready 4-m diameter telescope called LAMP (to be described later). It was never launched, but demonstrated a realistic capability. Another piece of information comes courtesy of the American Federation of Scientists, which has posited that a spy satellite launched in 1995 (part of a programme called Improved Crystal) is a KH-12 telescope with a diameter of 4.5 m. The last point to note is that classified military space technology tends to be 10–20 years ahead of that in the civilian world, and we know that the 6-m diameter James Webb Space Telescope is due for launch by NASA sometime around 2013. From all these pieces of information, we can reasonably assume that the military has a 6-m to 8-m diameter surveillance telescope in orbit now.

Taking the conservative value of 6 m, and putting all these numbers together, we can get a resolution of 2 cm. With such a resolution you could read vehicle licence plates and large newspaper headlines and possibly even recognise faces. Obviously this figure is merely an educated guess, but it is based on realistic estimates and simple physics. Still, there are some caveats. We have already discussed the blurring effects of turbulence when observing through the atmosphere, so you might think this would be a limitation. But looking down is not the same as looking up, since the turbulent layer is close to the ground. To see this, take a piece of bubble-wrap, and place it close to your eye. Looking at this page you will have trouble reading the text, due to the blurring effect of the plastic. With the bubble-wrap being so close to your eye, the effect is the same as having a poor-quality lens blurring your vision. This is also the case with ground-based telescopes looking at distant objects through atmospheric turbulence which is fairly close to the ground. Now place the bubble-wrap against the page and you will notice that you are still able to read the text. This is similar to the situation for orbital surveillance satellites looking down through the atmosphere. Of course, there is still some blurring, but the effect is much less than if the atmospheric distorting layers are close to the imaging optics.

Another factor we should consider is the true concept of resolution. The Rayleigh criterion is a useful rule of thumb, but it applies to a single image of a point-like object from a single location with a filled aperture. We have already discussed the fact that by using a distributed array of apertures it is possible to exceed the Rayleigh limit with a minor loss of contrast. In the case of surveillance, it may be acceptable to reduce contrast in exchange for improved resolution, so the possibility of using sparse apertures should be taken into account. More importantly, however, there are several techniques which can be used to improve resolution beyond the Rayleigh limit.

With a good deal of computing power, it is possible to use multiple images of the same object taken from slightly different perspectives to give a single image with 'super-resolution'. These techniques are well beyond the level of this text, but suffice it to say that they would make it possible to compensate for the small loss in resolution due to the turbulence, making our above estimate fairly good. Note that super-resolution techniques are not the same as the 'image enhancement' that is often shown in movies. You've seen it many times; someone is viewing an image from a surveillance camera and they are told to 'clean it up a bit' and magically you can see details that weren't apparent before. This is almost never the case in true surveillance imaging (or any imaging for that matter), where the pictures are always collected at the highest resolution the camera can manage. The post-processing of images mentioned here is computationally intensive and will only provide minor improvements in the quality of the final picture.

Advances in space imagery have also led to the creation of a new market in commercial satellite imagery. This began in the early 1990s with the French SPOT (Satellite Pour l'Observation de la Terre) satellites, and several companies have since launched orbital telescopes for the purpose of selling space-based surveillance images. Four companies: DigitalGlobe (US), Space Imaging (US), Orbimage (US) and ImageSat (Israel) each have their own satellites for hire and typical ground resolution is around 0.6 to 2 m, depending on the source and application. You can simply call up any of these companies and request a photo of a particular patch of the Earth. Several days later (depending on the orbital constraints), you will receive the image.

There are a huge number of uses for this sort of technology. Media outlets can purchase images of places of interest that are inaccessible to ground photographers. In 2001, when China captured a US EP-3E spy-plane, news outlets were quick to capitalise on this new capability. Space Imaging snapshots of the aircraft parked on a runway on Hainan Island were used extensively in news stories. More useful applications of this technology include urban development and planning, improved farming and crop analysis, disaster relief, environmental assessment and even archaeological research.

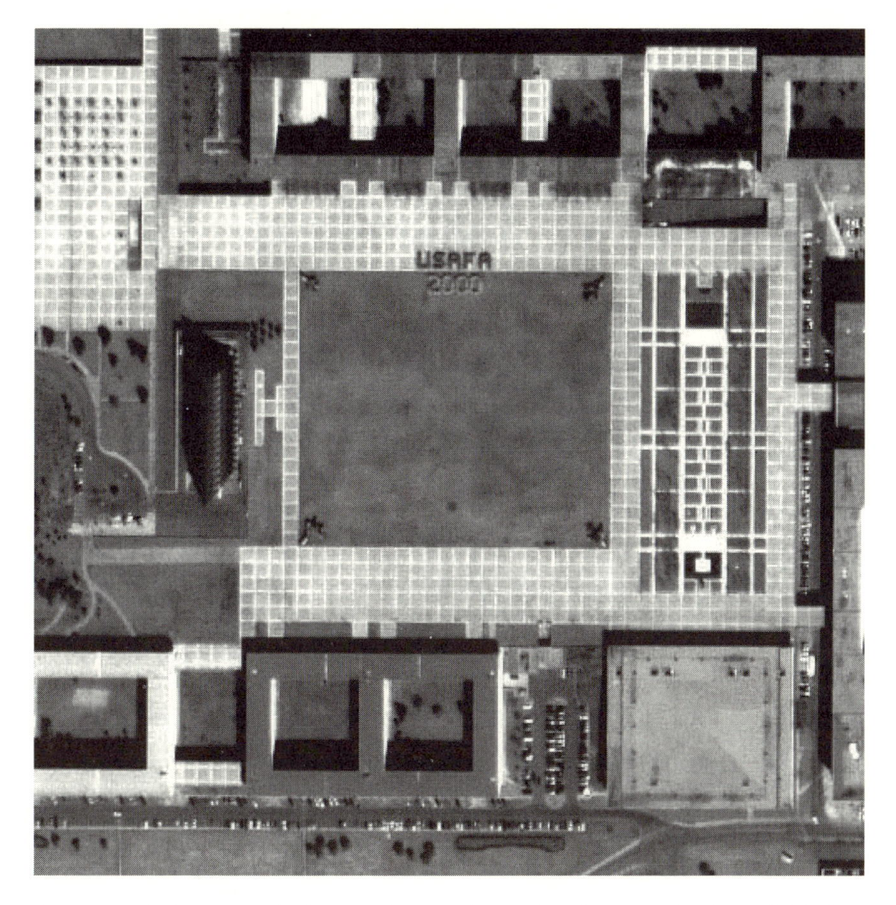

Figure 12.2: An image of cadets in formation at the US Air Force Academy taken by the IKONOS satellite in 2000. For scale, the squares on the Terrazzo are 8 m on a side. Courtesy: Space Imaging.

So can anyone use this technology? Well yes, within reason. The only real requirement is that you are not on the US 'Watch List' of known terrorists or undesirables. If you pass that test, you can pretty much purchase anything you want, so long as the satellite passes overhead at some point in its orbit. Take, for example, Groom Lake (Area 51). In 2000, the Federation of American Scientists acquired an image of the secret test facility which is generally the source of much debate amongst UFO devotees. The Federation then released this image to anyone who wanted to take a look. More than likely it came as something of a letdown to the conspiracy theorists, as it shows just what you'd expect from such a test facility: nondescript runways and hangars.

The true importance of the photo of Groom Lake lies not in what it shows, but in the fact that it exists at all. We have all seen the benefits which have come from such military technologies as GPS and the internet when they become available to the general public, and space-based surveillance technology will no doubt produce similar improvements in our lives. Today Google Earth is a service offered by the popular internet portal. With this freely available software you can visit anywhere on Earth using a massive database of surveillance images. Or to think of it another way: we can gain access to space-based images of our homes from the comfort of our homes. In 2005 an Italian computer programmer actually discovered the remains of a previously unknown ancient Roman villa simply by browsing images of his neighbourhood in Parma. At the moment, the highest-quality images are of the United States, but as this service gains more popularity, we can look forward to the day when we can view any part of the Earth at a resolution of half a metre.

Other surveillance methods

In spite of the increased priority towards space-based surveillance in the 1950s and 1960s, aerial methods still continued to be developed. In 1966, the SR-71 Blackbird was developed for the express purpose of flying beyond the reach of Soviet missiles using a combination of velocity and altitude. Flying at over Mach 3, at altitudes of up to 26 km (85,000 ft), Blackbird amply achieved this objective, even without its distinctive low radar cross-section. The SR-71 was designed for many surveillance missions from wide- and narrow-field optical imagery and radar sensing. The high-resolution telescopic cameras were capable of resolving ground details as small as 20 cm. The Blackbird was used for many surveillance missions throughout the Cold War, but was taken out of service in 1990. Today, however, aerial surveillance is going through something of a revival – a reconnaissance renaissance, you could say.

Compared with space-based telescopes, airborne systems have the benefit of longer dwell times and lower altitudes (which lead to improved resolution). Instead of large aircraft like the SR-71, though, smaller unmanned aerial vehicles (UAVs) are the preferred platform. The Global Hawk is one such UAV project, consisting of a remotely controllable aircraft with a 35-m wingspan capable of flying at an altitude of around 20 km (~60,000 ft) for up to 35 hours continuously. The bulbous fuselage (the length of a bus) is made of composites which give it a low radar cross-section, making it hard to detect. The precise types of sensors available for surveillance are a well-guarded secret, but it is reasonable to assume the fuselage could handle a 0.5-m diameter optical telescope that would permit imaging to a couple of centimetres. That is not all. The US Air Force is also considering a surveillance concept which is somewhere between the airborne and space-based systems. This so-called 'near-space' concept involves very large balloons (think über-zeppelins here) at extremely high altitudes. Capable of dwelling for days to months over a particular region, such a balloon could support massive payloads. A large telescope on such a balloon could provide real-time video feeds of the ground at centimetre resolution.

Surveillance is generally thought of in terms of looking down, but there is also the need to look up. Once nations began to put objects in orbit, it became vitally important to try to find out just what was being put there. These unknown objects could be weapons, scientific payloads or surveillance assets. By studying the nature of the launch and orbital parameters (with radar), it is often possible to infer the payload and its purpose. However, nothing beats getting a high-resolution image of the satellite for analysis. With this in mind, novel ground-based surveillance telescopes working in the infrared and visible spectrum are required. The United States Air Force has two such sites; the Air Force Maui Optical Surveillance and Supercomputing Site (AMOS) located in Hawaii, and Starfire Optical Range (SOR) located in Albuquerque, New Mexico. Both sites conduct a wide variety of operational and research programmes involving satellite and missile tracking/imaging, laser projection and communications along with the odd bit of astronomy. The sites were the platforms for the development of adaptive optics before the technique was declassified in 1990.

Whereas astronomical telescopes are generally steered at the leisurely sidereal rates of 15 degrees/hour, this is way too slow to track objects in low and medium altitude orbits around the Earth. As such, the 3.6-m Advanced Electro-Optical System (AEOS) at AMOS can move 18 degrees/second in azimuth and 5 degrees/second in altitude. To see the 70-tonne instrument as it whips around in a precisely controlled and whisper-quiet motion is really quite astounding. Equally impressive are the images it produces of satellites whizzing overhead. Figure 12.3 was taken by a smaller (1.5-m) telescope at AMOS. Note that this is an unclassified image showing much lower resolution than would actually be possible with the larger telescope.

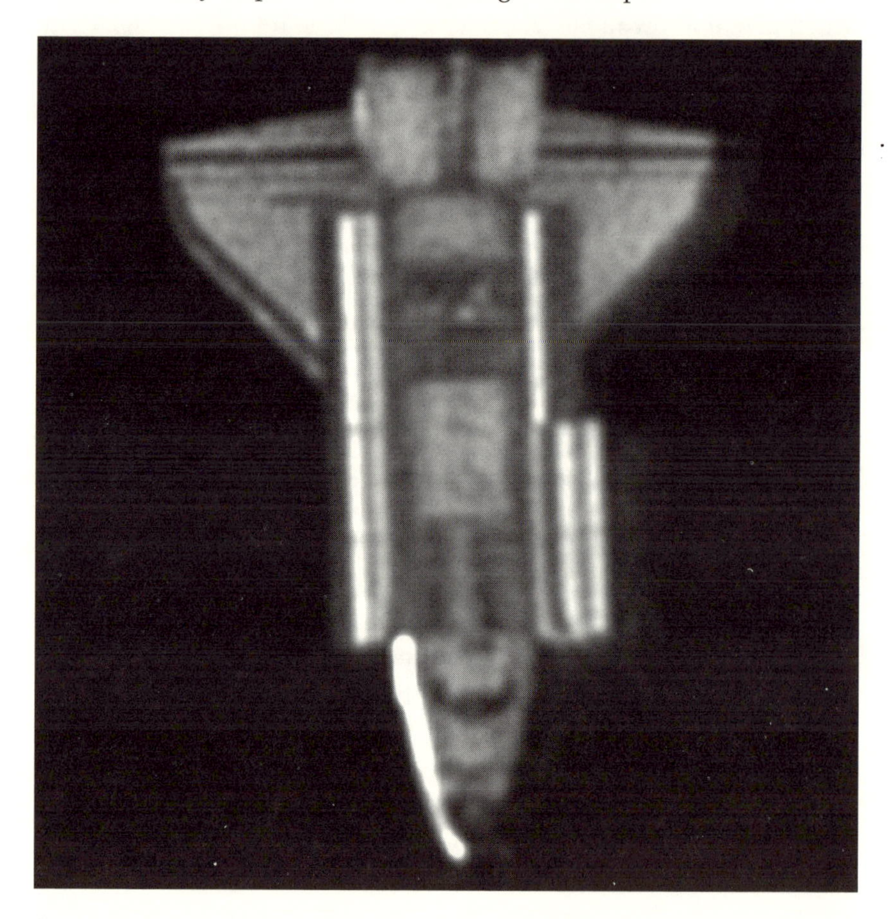

Figure 12.3: An image of the Space Shuttle *Columbia* taken from the ground on 28 January 2003 (STS-107). Four days later the orbiter was destroyed on re-entry. Image courtesy AMOS/USAF.

With the fast tracking of objects comes the problem in designing an enclosure. In conventional observatories, the dome is simply a protective covering with a gap in one side that rotates with the telescope. While this works fine for slow-tracking astronomical work, such large domes are difficult to move at the rates required by surveillance work while keeping vibrations low and ensuring no air movement (so as to avoid blurring the images). The solution for the two Air Force telescopes was to build collapsible domes. AEOS has a cylindrical dome which can be opened at the top and then hydraulically collapsed like a camping cup, to fully expose the telescope to the elements (Figure 12.4). This makes it possible to redirect the telescope rapidly while also giving access to viewpoints to the level of the horizon and below (something which astronomical facilities avoid).

Figure 12.4: The AMOS site on the top of Haleakala, Maui. The AEOS telescope is underneath the large collapsible dome on the left.

One last unusual aspect of the AEOS telescope is the beam path. Most modern astronomical telescopes have instruments mounted to the structure itself for stability. For example, they could be attached behind the primary (the Cassegrain focus) or light could be brought out through a hole in the yoke axis of rotation to large platforms on either side (the Nasmyth focus) of the telescope. The 3.6-m AEOS, however, has a Coudé path – where a number of small mirrors funnel the light from the telescope into the bowels of the building for analysis. For the AMOS facility, this has the advantage that light can be directed into one of many different rooms where different instruments are located. Some of these may be highly classified in nature, while others may house unclassified research projects. The drawback of a Coudé focus is that, when one looks down a long tube, the field of view is very small.

Of course, for the type of work done by this telescope it is not really such a problem.

Laser weapons

The transmission of lasers previously discussed was mostly passive in nature, but there is also the possibility of projecting beams for offensive purposes. In 1983, US President Ronald Reagan announced his directive for the military to pursue a strategic defence initiative (SDI). The so-called 'Star Wars' concept involved a plethora of space-based instruments designed to destroy intercontinental ballistic missiles launched from Eastern-bloc countries before they struck targets in the United States. One of the flagship devices was to be a space-based laser (SBL) which would be capable of destroying an Inter-Continental Ballistic Missile (ICBM) in transit. The idea of such 'ray guns' is attractive, but it is a very difficult task to put the idea into practice. Experiments conducted as part of SDI showed that a chemical laser was capable of producing the necessary multi-megaWatt output, but there was still the problem of focusing the laser onto a missile.

The requirement for the large telescope comes back (once again) to the resolution formula which determines the smallest focal spot we can get from a mirror, and as always, a bigger diameter is better. For this experiment, a telescope with a diameter of 4 m called LAMP (Large Aperture Mirror Project) was constructed. Shown in Plate 6, the telescope was a segmented primary mirror with special high-power reflective coatings. These coatings were designed to reflect as much light as possible and prevent the laser from damaging the mirror. The mirror you see in this picture was never launched into space as planned as funding for the SBL project was halted in 2000. However, another project did continue to be funded, in which a more modest laser is projected from an aircraft; the airborne laser (ABL).

ABL is a theatre-wide defensive system, which is to say that it is to be deployed in close proximity to the front lines. Instead of ICBMs, the targets are smaller, short-range missiles (think Scuds here). The ray-gun itself is a megaWatt-class chemical oxygen-iodine laser designed to fit (somewhat snugly) into the fuselage of a single Boeing 747-400. The laser is directed out of the nosecone using a 1.5-m diameter telescope. The telescope is designed to expand the beam and focus it on to a missile up to 300 km away. The idea is to destroy the rocket in boost phase, soon after it appears above the cloud layer. Note that the missiles will not be destroyed in the manner depicted in science-fiction tales – there will be no explosion in a shower of fireworks. Instead, the missile propellant is heated from the outside by the laser until it bursts its casing and tumbles uncontrollably back to Earth (preferably onto the guys who launched it).

Live tests of the first fully integrated ABL system will begin at the end of 2007, but there are still issues to be solved before it becomes operational. One of the major challenges is the effect of atmospheric turbulence (around the aircraft fuselage) on the beam, which reduces the focusing ability. In this case, the issue is complicated by the heat from the laser beam itself that is so intense that it generates further turbulence. As in the case of astronomical telescopes, the solution is to use adaptive optics. The dilemma for ABL developers is how to create a deformable mirror which is capable of operating under extremely high incident powers. While the mirrors are highly reflective, the fact that they are thin means that they are particularly susceptible to even minor heating effects. These and other issues are the natural result of pursuing any advanced research. As always though, the technologies developed in the course of this work are sure to find their way into future civilian telescopes.

Chapter 13

Non-traditional observatories

In this chapter we will consider some instruments which are not what most people would generally think of when the word 'telescopes' is mentioned. They all collect light to tell us about distant sources, but for the most part they don't fit into the traditional mould of the telescopes that have been discussed until now. Either they address niche research applications or they have a completely different manner of detecting light. In any event, it is precisely this uniqueness that makes them worth looking into in greater detail.

Liquid mirror telescopes

Isaac Newton was the first to propose constructing telescopes from mirrors instead of lenses as a means of removing chromatic aberration. The problem then became how to create such a mirror with high reflective properties. Newton made his first telescope from a highly polished piece of copper, but also suggested other possibilities including silver, brass, lead or speculum. Newton further thought of using a liquid metal such as mercury for such a mirror. In this case, the parabolic shape would be achieved by spinning a bowl of such a liquid. The idea was left largely undeveloped until the nineteenth century when several different people attempted to build liquid mirror telescopes – none of them with much success. This all changed in 1872 when Henry Skey made the first working device, in New Zealand. The mirror was 35 cm in diameter and produced an acceptable focus which could be altered in distance from the mirror by changing the rotational speed of the bowl of mercury.

The concept is quite beautiful in its simplicity. A rotating bath of fluid is subject to two major forces (gravity and centripetal). The downwards force of gravity acts equally everywhere over the liquid and keeps it in the container. Meanwhile the centripetal force acts radially, increasing in strength towards the edge. The combined result is that the fluid will move from the middle to the edges to create a perfectly parabolic surface. While the mathematical description is simple, the engineering realities are more complex. To begin with, the focal length of the mirror depends on the rotation speed of the liquid. In order to ensure that our images remain in focus, the rotation rate must be kept constant to a high degree of precision. The other problem is that the slightest of vibrations in the mechanical system will be transferred to the surface of the fluid, thus creating ripples. These ripples are just like imperfections in any mirror surface, and they can ruin the image quality unless kept to less than an eighth of a wave in size. It was in part due to these issues that liquid mirrors took so long to be successfully demonstrated.

In 1982, Emmanuel Borra at the Laval University in Quebec, Canada showed that a liquid mercury telescope with high optical quality was quite conceivable with modern technological advances. Over the next decade, both his group and a separate Canadian group under the direction of Paul Hickson, at the University of British Columbia, began fabricating and improving these mirrors. Some of their advances included:

1. using high-precision drives to ensure a stable rotation speed
2. using precision air bearings to reduce vibrations and maintain a perfectly balanced surface
3. fabricating epoxy containers of roughly parabolic shape to reduce the mass of mercury required and reduce surface ripple

One of the most important achievements of these groups was to demonstrate that these mirrors are quite safe to operate. Mercury is toxic and can accumulate in the body over a period of years. Although it is a metal, some mercury will evaporate into the air, even at room temperature. However, it so happens that the surface of the spinning mirror rapidly reacts with oxygen in the air and develops a thin top layer of mercuric oxide. This barrier then prevents further reaction and reduces the amount of mercury released into the atmosphere to well below harmful levels.

Another effect that initially caused some concern is Coriolis force. This force acts on objects which are travelling at an angle to a rotating inertial frame. It is an extremely weak force and very rarely observed in everyday life. However, on a planetary scale it can be significant, and on Earth is responsible for the generation of hurricanes and cyclones. The geometry of the effect is such that these systems rotate in opposite directions in the Northern and Southern hemispheres, but contrary to popular belief, it is far too weak to cause liquid to spiral around a drain differently on either side of the equator. Even in a huge spinning bath of mercury, the effect of Coriolis force is tiny. However, telescope mirrors are extremely sensitive to the slightest deformation of the liquid surface, so Coriolis force would be large enough to affect image quality in mirrors greater than 2 m in diameter. Quite fortuitously, then, it turns out that nearly all of the Coriolis error can be removed simply by applying a tiny tilt to the liquid mirror (north or south, depending on which hemisphere you are in).

Liquid mirror telescopes are limited to viewing objects near zenith. This means that astronomers are restricted to targets which happen to wander through the field of view at any given time. Objects can be revisited, but you have to wait a day. The inability to track objects also means that stellar objects will seem to streak across the image plane. When using a CCD to take the images, we can use a process called time-delay integration (or bin-shifting) to compensate for this motion. This involves shifting the light in columns of pixels to the adjacent column at precisely the rate at which the stellar objects move over the image plane. While this makes it possible to produce images free of motion blurring, the total integration time is only a few minutes (the time it takes for the star to cross the entire CCD). This is significantly less than the hours-long integration times that astronomers are accustomed to at conventional observatories. The result is that liquid mirror telescopes tend to be used for particular niche astronomical projects such as cataloguing and sky surveys.

Perhaps the best application of liquid mirror telescopes is atmospheric lidar. For the most part these sorts of studies involve probing a column of atmosphere using a laser. In most cases, straight up is as good a direction as any other, so the lack of telescope tilt is not a problem. Lidar researchers are accustomed to collecting weakly scattered light signals with small commercial telescopes, so the ability to get tens or hundreds of times larger signals for little added expense is very appealing. This was the reasoning behind the construction of a liquid mirror telescope for the University of California research station outside Fairbanks, Alaska. The High Power Auroral Stimulation (HIPAS) experiment conducted there involves stimulating the ionosphere with a megaWatt ultra-low frequency radio transmitter. The effects of this stimulation are observed by many methods, including a lidar system. In this case, a multi-megaJoule pulsed laser is projected into the sky to monitor the state of the ionosphere. The weak light which returns to Earth is then gathered by a liquid mirror 2.7 m in diameter.

There has been a gradual increase in aperture sizes for liquid mirrors over the last 20 years. In 2003, the Large Zenith Telescope (LZT) was commissioned at a mountainous site an hour's drive east of Vancouver, Canada. Built as a collaborative effort by Canadian, French and American universities, the main mirror is an impressive 6 m in diameter, making it the third-largest optical telescope in North America. Even so, at a cost of a mere $300,000 its price is around one-hundredth that of a conventional telescope. While the telescope is impressive enough to accountants, the engineering issues are startling.

Even though the layer of mercury is only around 1 mm thick, the 30 litres of it weighs about 400 kg. With the support container, struts, secondary and camera, the final mass is around 2000 kg. This massive structure is then precisely balanced and rotated at a stately 8.5 seconds for a complete revolution, to an accuracy of 1 part in a million. At a brisk walk you can keep up with a point on the rim. This of course leads to the problem of stationary air producing ripples in the moving liquid surface. We are used to ripples being caused the other way around but it is essentially the same problem. The solution turned out to be relatively straightforward – simply clamping a thin, stretched film of Mylar over the top of the mirror surface.

The images produced by this telescope have been quite spectacular. A photo of the mirror (spinning!) is shown in Figure 13.1. The telescope is run 24 hours a day, even when it is not being used for observations. As well as serving as a test-bed for further development, this instrument is conducting sky surveys and photometric studies of all sorts of objects.

Figure 13.1: The 6-m Large Zenith Telescope. The bowl of mercury is spinning, but the mirror shows no signs of it; there are no ripples or waves disrupting the perfectly smooth surface. Courtesy: Paul Hickson, University of British Columbia.

You would think that a 6-m mirror would be enough to make any astronomer happy. But telescopes are like a drug, with the only difference being that the cravings are for a larger aperture. The next generation on the drawing board is a mirror 8 m in diameter to be located on a mountain in Chile. The project has the catchy name of ALPACA (Advanced Liquid-mirror Probe for Astrophysics, Cosmology and Asteroids), and will be one of the largest telescopes in the world. But even that isn't enough. There are severe engineering constraints on creating a single liquid mirror telescope with a significantly larger aperture than ALPACA, but there is the possibility of combining several smaller telescopes for a large effective aperture. The current design for the Large Liquid-Aperture Mirror Array (LAMA) has 18 individual apertures: each one 10 m in diameter, and all closely packed with a circular patch of ground a little over 50 m in diameter. We have already discussed the difficulties in creating phased interferometric arrays, but this project has the added challenge of rotating liquid mirrors – something guaranteed to keep many a graduate student occupied. The numbers are impressive, but so is the estimated price-tag of $100 million, which would barely buy you a single Keck or Gemini telescope.

Solar telescopes

In the third century BCE, Archimedes is said to have suggested a method to protect Syracuse (in Sicily) from an invading navy. All that was required was to get a few hundred people with mirrors to surround the harbour and reflect sunlight onto the same enemy ship. The combined solar power would be enough to ignite the wood. This is probably the first example of directed energy weaponry ever discussed, and with enough people it could conceivably work. Even with a small magnifying glass, a delinquent child can cook an ant in seconds by focusing only a small amount of sunlight. On Earth, solar power amounts to a little over a kiloWatt per square metre, mostly in the optical and infrared, so a large telescope mirror could easily collect and focus enough light to melt steel. Of course, some astronomers want to look at the Sun with telescopes, so they need to find ways to avoid the inconvenient destructive effects of such observations.

Detailed observations of the Sun have allowed us to gain a better understanding of its surface activity and from this infer the processes taking place inside. Many observations of the Sun before the invention of the telescope were made with the naked eye and camera obscura (pinhole camera). With the advent of the telescope in 1609, magnified images showed such unexpected phenomena as the presence of imperfections on the solar disk. By timing the passage of these sunspots across the face of the Sun, it was deduced that the equatorial rotation rate was about 25 days. These spots are often many times larger than the Earth and appear black only because they are a mere 5000 K compared to the much hotter (and brighter) 6000 K of the surrounding photosphere.

Careful observations have indicated that powerful magnetic fields play a large role in the formation and development of sunspots. We also know that the number of spots seen at any given time varies over an eleven-year period. The maximum coincides with a period of increased solar activity which can cause radiation problems for satellites, electrical grids and astronauts. Oddly enough, from 1645 to 1715, soon after telescope observations began, the Sun entered a phase called the Maunder minimum. During this period there was a paucity of sunspots – an event that coincided with drops in temperatures in the Northern Hemisphere known as the Little Ice Age. Only through more careful observations can we hope to gain a better understanding of the solar environment and improve our forecasting of future activity. But beyond the direct relationship it has on our everyday lives, the Sun also represents a singular example from which we can directly test models of stellar dynamics. In turn, these models can help us infer much about other stars in the universe.

In spite of the abundance of light, a lack of which is the problem for conventional astronomy, solar astronomy is difficult because of the huge amounts of turbulence present. We have already seen that atmospheric turbulence can distort astronomical images at night, but solar telescopes have the added problem that the focused light is itself intense enough to cause local heating of the air. It gets worse still; there is the problem of dumping so much energy on to the optics inside the observatory. Under such extreme heating conditions, even large optical components will change shape slightly, which will distort images. To deal with these issues, solar astronomers have devised several solutions.

A solar telescope typically consists of one (heliostat or siderostat) or two (coelostat) flat mirrors which track the Sun and continuously direct the light onto a fixed primary mirror, usually a long distance away. Most solar telescopes have these flat mirrors mounted at the top of a high tower to get above as much ground turbulence as possible. The light from the flat mirrors is directed down through a structure which is painted white, to reduce the heating of the walls and the enclosed air. This distinctive characteristic of solar observatories is noticeable in the photos shown in Figure 13.2. The light can also be directed through large tubes filled with argon gas or even a vacuum, to further reduce absorption and heating.

Figure 13.2: From left: the Richard B. Dunn Solar Telescope; the Swedish Solar Telescope; the 150-foot tower at Mt Wilson. Courtesy: NSO/AURA/NSE, Royal Swedish Academy of Sciences, and Mt Wilson Institute.

A typical solar telescope has a primary mirror with an extremely long focal length. The Sun subtends an angle of half a degree, and so will the image formed by the primary. This means that the initial image formed by this mirror will be quite large, even before further magnification. With this in mind, solar telescopes often don't even use a secondary – there is simply no need for the higher magnification. Instead, the image formed by the long focal length primary is directed straight onto the camera. For the Dunn Solar Telescope, the primary mirror sits as far underground as the top of the tower above it. The focal length of the primary mirror is around 55 m, so the size of the Sun's image (back at ground level) is almost half a metre in diameter. It is truly an impressive sight to see the disk of the Sun projected onto a screen at this size.

Adaptive optics has created a change in the way solar astronomers conduct their observations. As with nighttime astronomy, deformable mirrors placed in the path of the light can be used to correct for atmospheric turbulence and heating effects on the mirrors. With conventional telescopes, the procedure is fairly straightforward – look at the blurred image of star and adjust the deformable mirror until it looks like a point. The star can be either a natural or an artificial laser guide within the field of view. When looking at the Sun, however, neither of these options can be used, since the light from the Sun is much brighter than a guide star could ever hope to be and any light from the guide star would be lost in the glare. Instead, a slightly more complex technique has to be used which relies on the fine structure on the Sun. There is no doubting the results, as can be seen in the image below (Figure 13.3).

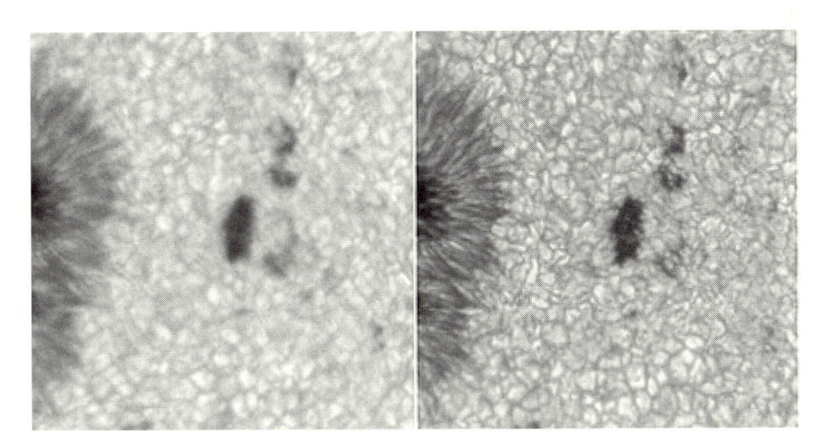

Figure 13.3: Images of the Sun without the use of adaptive optics (left) and with adaptive optics (right). The size of each box is approximately two and a half times the diameter of the Earth. Courtesy: NSO/AURA/NSF.

Until these recent successes with adaptive optics, solar astronomers had little reason to contemplate really large primaries. If you are limited by atmospheric turbulence, the only advantage to a larger aperture is in gathering more photons for brighter images – something that is never a problem with the Sun. But with adaptive optics, larger mirrors begin to make sense. With this in mind, a consortium of agencies has formed to build the Advanced Technology Solar Telescope (ATST). This telescope is to have a primary 4 m in diameter and will be vastly different to any existing instruments. To begin with, the primary has a very short focal length, which means the light comes to a very small, bright image. At this point there is a cooled aperture which blocks off all but the region of the Sun we wish to observe. A concave secondary is then used to magnify the image. The overall configuration is a Gregorian, but tilted slightly to get the optics out of the path of the light.

The ATST will be a remarkable telescope. With the adaptive optics working at peak expected performance, it should be able to resolve down to 0.03 arcseconds. At the Earth-Sun distance (1 AU) this represents a 20-km feature size on the Sun. For comparison, Plate 7 shows the current state of the art from the Swedish Solar Telescope showing details as small as 100 km. Looking at this image it is hard to imagine that the size of the sunspot is about the same as the diameter of the Earth. Even more impressive are the movies that have been taken. Unlike other celestial observations, the exposures taken through a solar telescope are extremely short. By combining many consecutive images, movies can be made to show solar dynamics. In such footage, the hair-like filaments are seen to stream into the sunspot, while the rest of the photosphere seems to boil. When the ATST sees first (sun)light in 2014, it is sure to provide us with even more impressive sights.

Seeing the invisible

The emphasis of this book has been on optical telescopes, as the extension into other parts of the spectrum would constitute an overwhelming amount of subject matter. However, it is possible that we can collect light signals on Earth that are the result of secondary effects from higher-energy radiation such as gamma-rays. Gamma-rays are the highest energy form of electromagnetic radiation and are produced in only the most violent processes. Some examples of these include supernovae, coalescing star systems, pulsars, black hole accretion disks and gamma-ray bursters. The latter are cataclysmic events, the cause of which is still something of a mystery. There are a few sources of gamma-rays to be found on Earth, including radioactive sources, high-energy particle accelerators and nuclear explosions, but the quantities generated are small.

Gamma-ray energies can exceed 10^{20} electron-volts. We tend not to deal with this unit of energy in everyday life, so it perhaps requires some clarification (or a short detour to Appendix B). A single photon of light tends to have an energy of a few eV, and a regular 50-Watt flashlight emits about 10,000,000,000,000,000,000 photons of light per second. By comparison, a single gamma-ray photon can have the same energy as all these visible photons combined. This should also go some way towards explaining why gamma-rays have such a destructive effect on human tissue. It is for this reason that gamma-ray sources can be used to kill cancers in radiation therapy.

The problem with detecting gamma-rays is a consequence of their high energy – they have a high penetrating power. That means that they tend to pass deep into or straight through even the densest material before they interact with an atom. As such it is not possible to construct lenses or mirrors to focus the light, or produce a high-resolution image, of a distant gamma-ray source. Even if imaging could be achieved in the conventional manner, the ephemeral rays would pass straight through most detectors. These problems are compounded as energies increase, so most gamma-ray telescopes are designed only to 'see' at the lower energy range of the spectrum (10 MeV or less). At energies a million times higher (1 TeV and beyond), it is extremely difficult to collect an appreciable number of gamma-rays in even the thickest and most sensitive detector. One solution is to use the entire atmosphere as the absorbing material.

When a gamma-ray strikes the Earth's atmosphere it can create showers of charged sub-atomic particles, some of which are travelling faster than the local velocity of light.[34] In much the same way as an object travelling faster than the speed of sound in air creates a sonic boom, the particles will create an optical equivalent: a flash of electromagnetic radiation (mostly in the blue and ultraviolet) called Cherenkov radiation. The burst of light is emitted in a cone which is centred along the direction of the original gamma-ray. By detecting the burst of light, it is possible to put together a picture of where the parent gamma-ray originated. This is the principle behind the High Energy Stereoscopic System (HESS) run by a collaboration of European and African scientists and based in Namibia. It is aptly named, since

Victor Hess was awarded the 1936 Nobel Prize for his discovery of this cosmic radiation. The HESS 'observatory' actually consists of four 13-m diameter telescopes placed at the corners of a square 120 m on a side. The telescopes point in the same direction and work in concert to positively identify and characterise faint flashes of Cherenkov radiation.

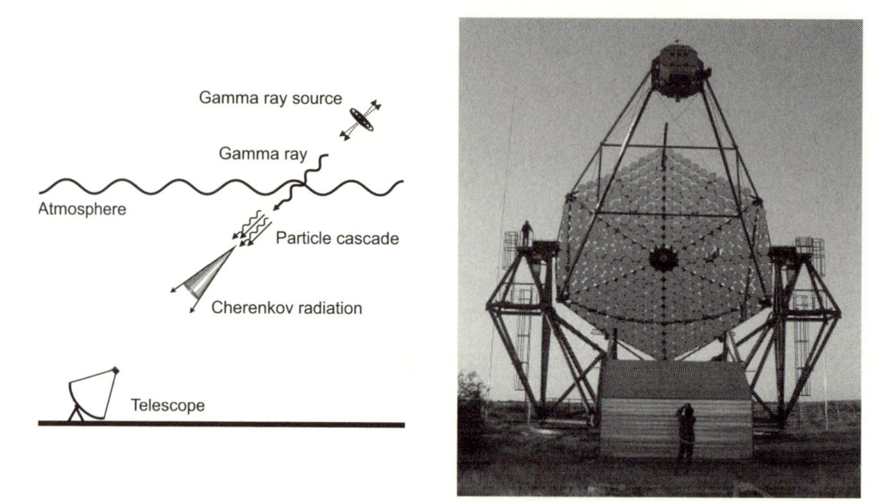

Figure 13.4: An artist's conception of how the HESS detects the Cherenkov light from a gamma-ray source (left), and a photo of one of the individual receivers (right).

As you'd expect for such a big project, the layout of the four telescopes was chosen quite carefully. The density of the atmosphere increases exponentially as the gamma-ray heads towards the ground, so the further it travels the more likely it is that it will interact with a molecule of air. Beyond a certain point, however, the number of showers begins to decrease as most gamma-rays have already collided with an air molecule. The result is that, while a burst of Cherenkov light could be generated at any height, they tend to preferentially occur at an altitude of around 10 km. Given the geometry of the cone, we can usually expect a Cherenkov burst from this height to illuminate an area some 250 m in diameter on the ground. For this reason, the HESS telescopes are positioned in such a way as to maximise the likelihood that the cone of Cherenkov radiation will intersect at least two telescopes. It should be mentioned here that while these telescopes have a larger aperture than traditional optical observatories, this is only possible because they have much lower quality optics and lower resolution. In essence, these telescopes are designed for light collecting with meagre angular resolution rather than for diffraction-limited imaging at optical wavelengths.

Using the collected light signals, it is possible to build up an image of the source that emitted the gamma-rays. This is not as easy as it would first appear. To begin with, the detectors must be quite sensitive. The flashes of light are very weak, which should be evident from the fact that you have never seen them. While the original shower may consist of ten thousand or more photons, these are spread over a wide area on the ground. Even with the large-diameter mirrors used to collect the light, only a few hundred or so are actually detected in each burst. In fact the detectors themselves are so sensitive that observations cannot be made whenever the Moon is above the horizon, as even scattered moonlight swamps the target signal. Fortunately the background noise is spread out over a much longer time than the duration of the Cherenkov signal so it can be averaged out. Alternatively, by comparing the timing of events from multiple telescopes, the physicists can positively identify a flash in each one as being from the same cone of light, and not from two or more separate background events that just happen to occur at nearly the same time.

It is important to note that the showers are almost never aligned in the direction that the telescopes are pointing. This means that a flash usually appears off-axis in the field of view of the telescopes, and so, rather than seeing nice point in the image plane, we get an elliptical stripe. The stripe is aligned along the axis of the Cherenkov cone, so by combining the stripes from multiple detectors, it is possible to calculate the point in the sky from which the gamma-ray originated. The last step is to combine observations of countless flashes over many nights to build up a picture of the sources which created the gamma-rays. With care, the final resolution of the telescope in imaging a gamma-ray source is a few arcminutes – not much worse than the human eye.

In late 2004, the HESS group published an image of the remnants of a supernova (Plate 8). It is emitting a large amount of high-energy radiation as a result of the expanding shock wave striking the interstellar medium. The image is constructed from parent gamma-rays in the 0.1–10 TeV energy range and it required around 18 hours of viewing to build up an appreciable number of events to give the false-colour image shown. The supernova occurred in 393 CE (or more precisely, the light from it reached Earth in that year), and Chinese accounts report that it was a memorable sight. Over the last 1600 years it has faded to nothing, as can be seen (or rather, not seen) in the image of the same portion of sky taken in the visible (Figure 13.5). To give you an idea of scale, this image is around twice the diameter of the Moon.

Figure 13.5: An image of the same star field as in Plate 8, but taken at visible wavelengths. It is presented in negative to the right.

It is clear from this example that if we could 'see' gamma-rays, the sky would look remarkably different than it does in the visible. In this case, the supernova remnant would be a spectacular sight in the night sky. Along with a number of previously invisible objects, we would also see occasional bright flashes called gamma-ray bursters. In all, we would see many of the most energetic and most violent processes taking place in the universe. Short of quickly evolving gamma-ray vision then, Cherenkov radiation telescopes offer us a new window on the heavens. But as it turns out, we can use another unusual type of optical telescope to 'see' many of the same cataclysmic events.

Gravitational wave observatories

On either side of the United States there are two L-shaped tubes some 4 km in length, which enclose an instrument trying to detect minute distortions in the fabric of space-time itself. The long pipes encompass Michelson interferometers of such sensitivity that they are defining the very limits of what is technologically possible. Ultimately the data collected by each interferometer will be combined with that from other such gravitational wave detectors in Japan, Europe and perhaps Australia. The result should be the first experimental detection of a source of gravitation radiation – something predicted by Einstein's General Theory of Relativity, but never seen directly. The combination of interferometers will form arguably the largest observatory in existence to detect, somewhat counter-intuitively, the smallest of physical effects known. In order to understand the operation of this unique 'observatory', we must start with a little background on Einstein's theories and the effects they predicted.

In 1905, Einstein published five seminal papers which dramatically changed our view of the universe. Four of them are unfairly given very little press, but I will not be rectifying this oversight as they have little to do with this topic. The fifth, however, is perhaps his best-known work, on the Special Theory of Relativity. Here I will merely describe the basics required to understand later developments and help you to deal with all but the weightiest conversations at your average dinner party. We begin by casting our minds back to the turn of the twentieth century when physicists were struggling with the problem of the æther. As we discussed in Chapter 7, the æther was initially a putative requirement for the transmission of light waves. When Michelson and Morley demonstrated the absence of the æther, the problem of how electromagnetic radiation propagated through space returned. In one of his other papers, while addressing a completely separate phenomenon, Einstein postulated that light could come in quanta (or bundles) called photons.[35] As a bonus, this also solved the problem of the æther – it simply wasn't necessary, since particles can travel in a vacuum.

So far so good, but the problem then becomes one of a frame of reference. One reason physicists liked the æther is that it served as a frame of reference for all objects in the universe. In the same way you might define your location by longitude and latitude on the fixed sphere of the Earth, astrophysicists wanted an æthereal framework in space which they could use to set positions of all objects. Similarly, there was a desire to also have a temporal frame of reference. Think of this as some single clock which can be used as a time standard to which everyone anywhere can agree. The æther would have served this purpose, but when Einstein took this away, he had to find a solution to the frame of reference. In his paper on Special Relativity (SR), Einstein made two postulates:

1. Everyone will measure the velocity of light to be the same value irrespective of their relative velocities.
2. The laws of physics are the same in any inertial frame of reference.

In setting these axioms, Einstein effectively removed both absolute spatial and temporal frames of reference. Two observers may not agree on where or when they are, but if they measure the speed of light they will get the same value of 300,000 km/s. While Special Relativity could be verified by many experiments, it was incomplete in that it only applied to reference frames having a constant velocity.[36] Einstein spent the next ten years devising his General Theory of Relativity (GR), which 'generalised' relativistic effects to describe objects under acceleration and thus included the effects of gravitation. Einstein began with the postulate of the equivalence of inertial and gravitational mass: i.e. that you cannot tell the difference between gravitational or regular acceleration. He then went on to demonstrate how gravitational acceleration could be viewed as a direct result of the curvature of space-time. In simple terms, all masses will distort the shape of space-time. Gravitational effects are then merely the manifestation of how masses react to this curvature. In Figure 13.6 we can see an abstract presentation of how the Earth might deform space-time. It is easy to see from this image just how a mass placed on this curved surface would naturally be attracted to the Earth. There is much more to GR, but this is sufficient to explain most of the following material.

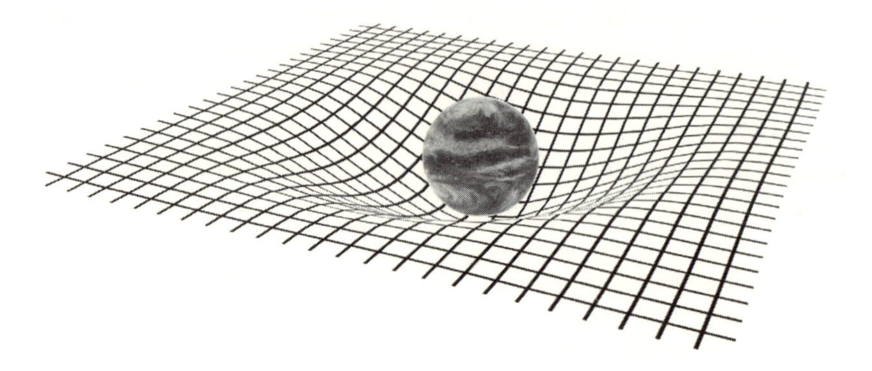

Figure 13.6: An artist's conception of the curvature of space-time due to a mass (in a two-dimensional manner).

Once again, physicists have worked long and hard to verify/disprove GR in numerous ways, using both man-made experiments and astrophysical observations. In all cases so far, GR has held up to the scrutiny – in many cases to a precision of parts in a billion or better. One of the first tests was made by an Englishman, Arthur Eddington, who examined one of the predictions that light would take a curved path when passing close to a massive object. If this were true, stars viewed near the edge of the Sun would have shifted positions compared with the same field viewed away from the Sun's influence. Stars are all but impossible to see with the Sun close to them, but Eddington made the observations during the solar eclipse of 1919. He claimed to have observed the predicted shift in stellar positions, thus verifying the theory. Later examination of his data suggests that it was debatable that he could have seen the small shift, but many subsequent observations have indeed verified the expected effect.

General Relativity also predicts that accelerating masses will radiate energy in the form of gravitational waves. These waves are somewhat analogous to the way in which an object moving on the surface of a lake can send out water waves in all directions. In effect, gravitational waves cause a rippled stretching and shrinking of space-time itself. Since gravity is a very feeble force, even the gravitational waves emitted by very massive objects under violent accelerations are extremely hard to detect. Until now, we have only been able to infer the existence of gravitational waves by observing astrophysical phenomena. For example, accurate timings of the orbital period of binary neutron stars have shown a minute increase in speed (of the order of microseconds per year) in precise agreement with GR predictions on such binaries radiating gravitational energy. Still, a direct observation of these waves has yet to be made.

Two observatories in the United Sates are attempting to make a direct detection of gravitational waves as they pass through the Earth. The experiment is called the Laser Interferometer Gravitational wave Observatory (LIGO), and is the culmination of years of design and construction. In this case the 'telescopes' are a pair of the world's most sensitive Michelson interferometers, each with arms 4 km in length. If a gravitational wave passes by, it will alternately stretch and contract the arms by an infinitesimal amount. The predicted stretching amounts to less than 10^{-18} m – around one ten-millionth of the diameter of a hydrogen atom. The fact that these numbers are so hard to put into everyday terms is in itself an indication of just how extraordinarily difficult this type of measurement is. A Michelson interferometer can measure very small distances, and in my lab I could carefully construct a device incorporating an expensive laser and high-quality optics on a stabilised platform to measure distances to a precision of a hundredth of a wave or around 5 nm. However, this is still 10 billion times larger than that required for detecting a motion induced by typical gravitational waves. In order to achieve the required resolution then, dramatic improvements are required.

To begin with, there are the lengths of the interferometer arms. The effect of a passing gravitational wave is to cause a strain, or stretching of the space. Imagine a piece of rubber with 11 marks at a regular spacing of 1 mm (over a total of 10 mm). If we stretch the rubber to double the size, the separation between each tick mark will increase by only 1 mm. Meanwhile, the ticks at either end will now be 20 mm apart – an increase of 10 mm. A gravitational wave affects space in much the same way, so it helps to make the measurement over as large a distance as possible – hence the 4-km arms of the interferometer. But we need to make sure that the stretching we are measuring is not from other effects. For example, if the light passes through even the slightest whiff of air, there will be a massive lengthening or shortening of the optical paths. As a result, the entire interferometer must be enclosed in a chamber under an ultra-high vacuum.

Next comes the laser, which is a one-of-a-kind system delivering a high-power infrared beam. The wavelength and brightness are controlled to ridiculous precision with a multitude of electronic feedback systems before being sent into the interferometer. By now you should not be surprised to find that the mirrors in this interferometer are not the kind you see in your bathroom. They are 10-kg hockey-pucks of the finest glass polished to a flatness of around one thousandth of a wavelength of light and then coated to reflect 99.99995% the incident light. The whole purpose of the time and expense devoted to perfecting the mirrors is to ensure that they do not scatter or absorb any light. In fact, this low light allows us to bounce the light around each of the interferometer arms many times – building up the power inside the interferometer to around 6 kW.

Of course, all of this would be for nought if the optics were to move by the slightest amount. Acoustic noise and seismic vibrations are all sources of motion which would disturb one of the mirrors relative to the other. This could either mimic or completely swamp the signal we are trying to measure, so the mirrors must be isolated from the outside world. To do so, the mirrors are suspended by thin filaments from tables which themselves rest on a complex series of springs and dampening systems in large towers. Again, this is all done inside two vacuum tubes 1 m in diameter and 4 km long, shown in Figure 13.7. In fact, vibrations are one of the reasons that two LIGO interferometers were constructed on either sides of a continent. By comparing the signals from each, we can rule out sources of noise which are due to events in close proximity to either one. Only when both see the same signal can we be sure that it came from a distant source.

Figure 13.7: Photos of part of the LIGO interferometer in Livingston, Louisiana (left) and the mirror suspension system being positioned inside one of the arms of the interferometer (right). Courtesy: LIGO/NSF.

After all of this effort, the beams sent up each arm are recombined to form a null fringe – that is to say, a dark spot.[37] It may seem like a lot of work to end up with nothing, but there is a point to it. In the case of LIGO, the dark spot will be completely devoid of any light so long as the two arms of the interferometer maintain a fixed difference in length. If one of them grows or shrinks relative to the other, then the tiniest amount of light will appear. A gravitational wave will produce precisely this sort of change in the interferometer arms. But to be sure that any measured signal is not due to some local disturbance, the signals from the two LIGO interferometers are compared. Because gravitational waves are expected to travel at the speed of light, there may be a little time difference – up to 1/100th of a second – in the arrival time at the two observatories, which are 3000 km apart. The time delay will also help us narrow down the position of the source to somewhere in a circular band in the sky. With the help of other interferometers being constructed by European (VIRGO/GEO) and Japanese (TAMA) groups, we should be able to triangulate the precise location of the source. In this way, the consortium of gravitational wave detectors will act as a single observatory to identify and catalogue all sorts of cosmic phenomena.

The two LIGO detectors began their first uninterrupted year-long observation in December 2005. In its current configuration, LIGO should be capable of detecting two neutron stars spiralling into each other at a distance of 20 megaparsecs.[38] Although the statistics on the occurrence of these events is not certain, there are estimates that LIGO should have around a 50% chance of seeing an unambiguous signal within a year of operation. In order to improve the number of observed events, even better sensitivity is required. With this in mind, an upgrade called Advanced-LIGO is planned. In the upgrade, the sensitivity will improve by a factor of more than 10, and the detection rate by a factor of 1000 – leading to the detection of several gravitational wave signals every day. But all this will require higher laser powers, bigger and better mirrors and better isolation. The aim is to reach a sensitivity approaching one part in 10^{22}. To put this another way, if the 4 km arms were expanded to reach the Sun, the 'stretch' to be measured would still only be the size of an atom.

So what would a signal look like? Well, it really depends on the type of event that occurs. In the case of two small black holes spiralling into each other, the observable event will last less than a second, with oscillations occurring in the order of a hundredth of a second. The Earth-bound gravitational wave detectors are optimised to see this type of event, and will probably also observe rapidly spinning pulsars and a background static of cosmic gravitational noise. But just as conventional telescopes are designed to observe certain parts of the electromagnetic spectrum, gravitational waves also have a 'spectrum' and gravitational wave observatories are only able to 'see' over a certain part of it. Thus ground-based interferometers are blind to some sources of gravitational waves in the universe. These include the slow oscillations caused by supermassive black holes orbiting each other in the centre of distant galaxies. To detect these, another space-based interferometer (called LISA) is being planned for operation in about 2015. This one is to consist of light bounced around in a triangular path between three mirrors – each 5 million km apart! The experiments are ambitious, but the payoff is that when they become operational a whole new universe will open up to us.

Chapter 14

Key discoveries

While telescopes were historically of great value for navigation and warfare, for the last four centuries they have also been put to work in observing our universe. The public support of these expensive instruments represents our basic desire to know just what is 'out there'. But we are also given a sense of wonder in seeing new and unimaginable sights brought to us by astronomers. Often these images are the result of determined observations into a particular phenomenon or target object. However, often the most impressive objects observed are simply the result of chance or of the advent of a newer, better telescope. There can be beauty in both. In this chapter, I have tried to highlight some less widely publicised images with a view of presenting a selection of oddities and unusual results which may not have gained much press.

The Solar System and Pluto

Telescopes have taken so many photos of our Solar System that we now think of its other planets as part of our family. We can all picture in our minds the rings of Saturn, Jupiter's Great Red Spot, or even the vibrant blue of Neptune's atmosphere. And yet none of these features is visible to the naked eye. The family portrait gallery began when the Russian *Luna* probe gave us our first images of the far side of the moon in 1959. It was found to have a completely different appearance to that of the side which constantly faces the Earth. The surface is covered in a multitude of craters, which now all have Russian names. Other probes from other countries have since made landings, orbits or fly-bys of every planet, the Moon and even some asteroids and comets. We have seen lightning on Jupiter, dust-devils and ice on Mars, aurorae on Saturn, geysers on Triton and volcanoes on Io. It seems the more closely we investigate any planet or moon, the more unexpected discoveries we make. In the case of Pluto, however, we know comparatively little.

Pluto was discovered by Clyde Tombaugh in 1930. The significance of his discovery is often overlooked, but it is worth noting that at the time only three people in history could claim to have discovered a planet in our Solar System. Tombaugh was born in Illinois in 1906 and developed an interest in astronomy at an early age. At the age of 20, with only a high-school education and self-taught in trigonometry and geometry, he constructed a reflector telescope with a diameter of 23 cm. He then sent images of the telescope and of observations made of Mars and Jupiter to Lowell Observatory in Flagstaff, Arizona where they were impressed enough to offer him a job. His initial work there consisted of searching for the cause of irregularities in the orbits of Uranus and Neptune that were thought to have been caused by the gravitational interaction with an as-yet undetected planet dubbed Planet X.

Tombaugh had the somewhat tedious job of taking twin images of large patches of the night sky, several nights apart. The photographs were then put in a machine called a blink comparator, which allowed the viewer to rapidly switch between views of each star field. Objects in the Solar System will become noticeable by the way in which they shift from one position to another against the fixed background star field. In February 1930, Tombaugh noticed a small point of light doing exactly this and calculations showed that this was an object, which he named Pluto, in distant orbit around the Sun. It later turned out, after the size and orbit of Pluto were measured, that it was way too small to have any significant effect on the orbits of Uranus or Neptune. In time it would be discovered that the calculated anomalous motions of these two planets were in fact mere artifacts, and there was no requirement for Planet X anyway.

As a side note here, it is worth mentioning the recent demotion of Pluto from a planet to a 'dwarf planet'. The issue was precipitated by the discovery, in July 2005, of a potential 'Tenth Planet' originally designated 2003 UB313 and since given the name Eris. It is about 5% larger than Pluto (and not much smaller than our Moon) while orbiting much further out in a highly inclined plane. Did this make it a new planet? The issue had previously been addressed by the International Astronomical Union (IAU) as they tried to reconcile the status of Pluto given the increasing number of similar objects being discovered beyond Neptune. Otherwise known as trans-Neptunian objects (TNOs), these bodies are most likely debris left over from the formation of the Solar System. Every year, more of these objects were being discovered, which began to make Pluto appear less and less like a 'conventional' planet.

The problem stemmed from the fact that there was no hard and fast rule as to what was, and what wasn't, a planet. The *Oxford English Dictionary* had a surprisingly vague definition to the effect that it is 'The name given to each of the heavenly bodies that revolve in approximately circular orbits round the Sun'. This definition did little to resolve the issue (as it seemed to add asteroids and a few comets to the list), so the solution was to find a clearer definition. The difficulty was in finding a set of criteria unique to planets, which is not as easy as you may at first think. For example, you might specify that a planet must have an atmosphere. All current planets have atmospheres, but so do some moons, and besides, the TNOs may have atmospheres too. Imposing a requirement that a planet has a moon is no good either, as Mercury and Venus don't have them, and some asteroids do. Anyway, as it turns out, Eris has a moon too (called Dysnomia). In 2006 the IAU decided to settle the matter with a complex (and still somewhat vague) definition that reclassified Pluto as a dwarf planet. It has caused a big stir, but is probably right given that a single change to textbooks now is simpler than changing them every year as possibly thousands more similar objects are discovered.

Pluto has a highly elliptical orbit around the Sun (the orbit is four times as long as it is wide) and at times it actually passes inside the orbit of Neptune. It orbits at between 30 and 50 AU from the Sun – a distance so large that the surface temperature is less than –220 °C. The rocky ball is only 2300 km in diameter and appears as a spot at a mere 14th magnitude – some 250,000 times fainter than Saturn. In 1978 James Christy discovered a large 'moon' orbiting Pluto which he named Charon. Charon is 1200 km in diameter and orbits Pluto at a distance of 19,400 km once every 6.4 days, often passing in front of its larger sibling, as viewed from Earth. It is a combination of the similarity in diameters between Pluto and its moon and the convenient geometry with Charon's orbit which has made it possible to take high-resolution images of Pluto's surface features.

With the Pluto/Charon system so small, it took several years before any telescope image was taken which actually showed two separate bodies. This number was increased to four in 2005 when an image taken by the Hubble Space Telescope showed a further two smaller moons named Nix and Hydra (Figure 14.1). We have discussed the resolution limit at great length, and by now the application and meaning of the resolution limit formula ($1.22\lambda/D$) should be well understood. If we now consider turning our telescopes towards Pluto, we can predict the smallest ultraviolet feature visible with the Hubble Space Telescope to be 250 km across. Given that Pluto has an estimated diameter of 2400 km, we cannot expect to make out any surface feature that is smaller than a tenth of its diameter. However, a clever idea was used to make it possible to circumvent this limit and actually make maps of the surface showing features five times smaller.

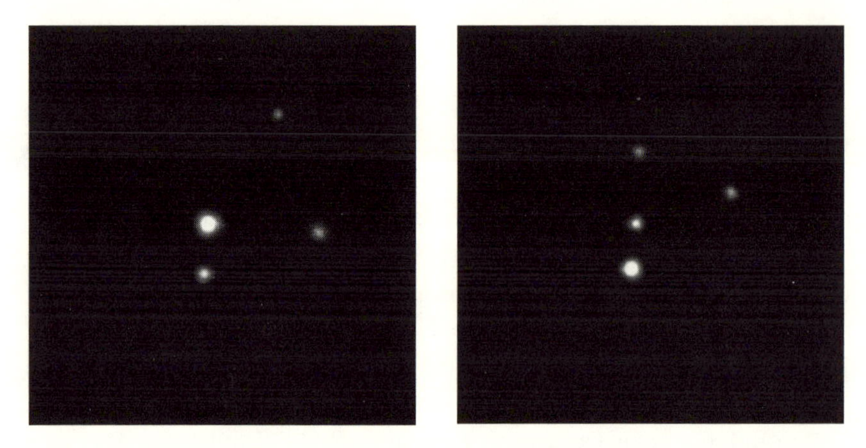

Figure 14.1: Pluto, Charon, Nix and Hydra are shown in two exposures taken three days apart. This image was produced by dramatically reducing the brightness of Pluto and Charon so that the two much dimmer bodies are visible. Courtesy: Hubble Space Telescope (STSCI/AURA).

In 1999, a group of American astronomers led by Eliot Young announced the results of an experiment which used the HST to produce a high resolution image of the surface of Pluto. They did this by making careful observations of brightness changes in Pluto as Charon passed in front. Imagine it this way: say we see the headlights of a car so far off in the distance that they appear to our eye as one single light. We then have someone walk in front of the car to alternately obscure one then the other light. From our vantage point we will notice a dimming and brightening of the 'single' point of light. Even though we cannot resolve the two individual headlights, we can see the brightening and dimming that implies their existence. In fact, if we knew the speed at which the person was moving, we could even calculate the separation of the lights.

By much the same principle, the astronomers used the edge of Charon as it passed in front of Pluto. For this to work, you have to precisely monitor the motion (position and orientation) of Pluto, Charon and the Sun while also measuring changes in brightness to a high degree of accuracy. Over many transits and with much number-crunching, they were able to build up a very detailed map of the surface of Pluto which is shown in Plate 9. Slight improvements over the years may improve this image somewhat, but we will probably have to wait until 2015, when NASA's *New Horizons* spacecraft arrives to reveal its full glory.

Comet Halley

Edmund Halley was a contemporary of Newton and, in fact, was largely responsible for Newton's publication of a theory of gravitation, *Principia*. Halley made a name for himself by completing a high-resolution catalogue of objects in the Southern hemisphere in 1678. In 1691 he suggested that timings of transits of Venus could be used to determine the distance from the Sun to the Earth and hence set the scale of the Solar System. Subsequent attempts to do so proved unsuccessful, but the theory is sound. In fact it was this idea that ultimately led to the 'discovery' of the east coast of Australia by Captain James Cook in 1770 when he was sent to the Southern Hemisphere to observe such a transit. Halley was the first to notice differences between the current positions of many catalogued objects to those recorded by ancient Greeks. He also correctly supposed that the stars themselves moved relative to each other as well as to the Earth.

In spite of these and many other contributions, he is still best known for his publication of *Synopsis Astronomia Cometicae* in 1705. In that work he used Newton's gravitational theory to surmise that the comets of 1456, 1531, 1607 and 1682 were in fact the same comet on an elliptical, 76-year orbit. In a marvellous demonstration of scientific method, he then suggested a test, making the prediction that the comet would reappear in 1758. Though he did not live to see it, the comet did return as predicted, and it has since returned three more times, the most recent being in 1986.

Comet Halley has an orbit that takes it 35 AU from the Sun. As of March 2003, it was 28 AU from the Sun, but will not reach its aphelion (most distant point in its orbit) until 2023. To say that the comet is dim would be an understatement. At magnitude 28.2, it is nearly a billion times fainter than the faintest object visible to the naked eye. In fact, it is too dim for any ground-based telescope to image it in a conventional manner. At the Very Large Telescope, any one of their 8-m telescopes taking a snapshot in the direction of Halley would show nothing at all. While we usually think of comets as impressive streaks of light, they really only become active as they enter the inner Solar System. As they pass close to the Sun, volatile compounds on their surface are vaporised to form the bright tails we are accustomed to. Meanwhile, in its current location at the outer reaches of the Solar System, Comet Halley is a dark, inactive, ball of rock and ice. This is not only because of the feeble sunlight far out in the Solar System, but also a result of the fact that the 10-km dirty snowball only reflects about 4% of the light it does receive.

At these sorts of light levels, any image taken by a telescope would be completely swamped by background noise. Some sources of this noise are the sky glow of molecules in the air, high-energy particles impacting the CCD and noise introduced while reading out the CCD. Even an image of a perfectly empty portion of the night sky would always show a very small amount of random lighting of pixels which can obscure a dim object. Not only is Comet Halley faint, but it is moving as well, so the weak light would be further dimmed by being spread over a line during a long exposure. Unless, of course, you know the precise orbit of the comet and can combine the many dim images taken while tracking the moving target. This is precisely what a group of astronomers led by Olivier Hainaut did. If this sounds easy, then let me break it down into the steps carried out.

The one factor working in the astronomer's favour is that precise tracking can ensure the object will always appear in the same place in the field of view. Thus we can take many images over a long period of time and add them up on top of one another (called integrating) to lower the effect of the noise. In order to understand this better, take a look at Figure 14.2. Here we have a 'signal' of 'strength' 2, while the random noise surrounding it can take any value between 0 and 3 (with an average of 1.5). In a single measurement the signal is lost in the noise, but if we combine many measurements, the signal will always be 2 each time, while the noise adds up to give an average value of 1.5. The plots for 1, 20 and 1000 observations show how a greater number of integrations increase the signal-to-noise ratio.

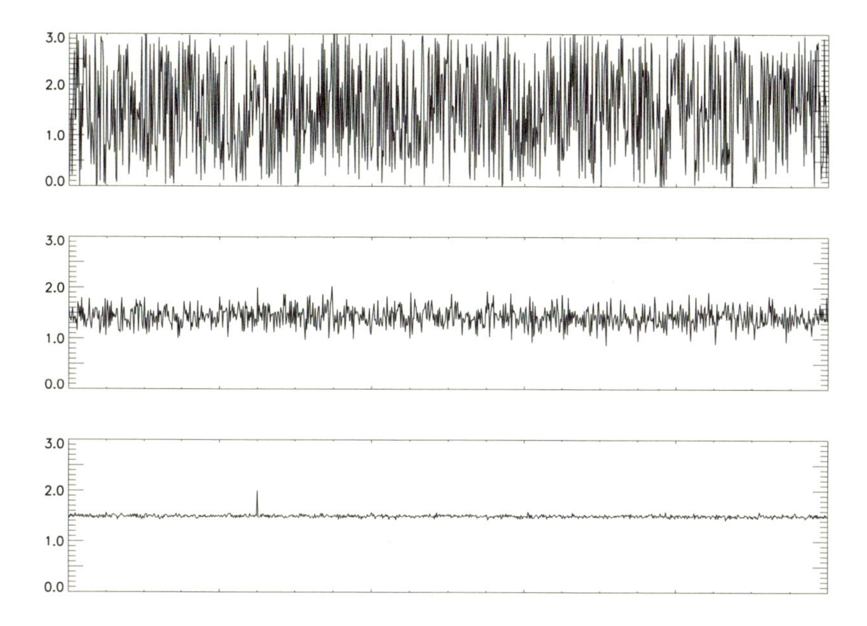

Figure 14.2: Signal to noise. By integrating data, a signal which is initially below the background noise (top) can be barely detected (middle). Increasing the number of integrations will produce even better results (bottom).

The same procedure was adopted by the group at the VLT. In this case, 81 images of the comet were taken over three nights by three of the four telescopes. Normally, when viewing faint objects, astronomers would take a single exposure, lasting an entire evening if possible. In this case, because the comet would move an appreciable distance in a single night, each exposure was only about six minutes long. The images were then 'stacked' on top of one another to give the final image shown in Figure 14.3, where the star trails have been removed for clarity. In all, only a mere 20,000 photons from the comet were collected to form this image. It is also interesting to note that these images, together with the images taken by the *Giotto* spacecraft which made a flyby of the comet nucleus in 1996, mean that Comet Halley has now been imaged by two European groups from distances of 600 and 4,200,000,000 km respectively.

Figure 14.3: The dim fuzzy spot in the centre is Comet Halley at a distance of 4 billion km from the Earth. The background noise is still evident even after the extensive image integration. Courtesy: ESO/VLT.

The first exo-solar planet

In 1995 a pair of Swiss astronomers, Mayor and Queloz, announced the discovery of a Jupiter-size planet orbiting another star. This was a major revelation, but the scientific community accepted it with some trepidation. After all, this wasn't the first time they'd heard such a claim and none of the previous 'discoveries' had held up upon closer scrutiny. This time would prove to be different. The planet was found to be orbiting 51 Pegasi, but couldn't actually be 'seen' as such. Rather, the astronomers used Doppler-shift measurements to notice its pull on the star and begin a new era in astronomy.

Perhaps the biggest of all questions we have as human beings is: 'Are we alone?' The average galaxy contains around 10 billion stars and there are around 100 billion galaxies in the universe. This means the number of stars in the universe might be comparable to the number of grains of sand on all the beaches of the entire world. Given all these star systems, it seems impossible that life exists only on Earth. The first step in approaching our question, then, is to find another planet orbiting around another star. For years astronomers had struggled with this problem as even a Jupiter-sized planet would appear billions of times dimmer than the star it orbits. The problem can be likened to trying to see a firefly buzzing next to a spotlight that is being shone in your eyes. Given these apparently insurmountable difficulties, astronomers decided to take another tack: spectroscopic measurements.

The process begins by observing the spectrum of the star over a long period with a very high-resolution spectrometer. If you recall, we can 'see' a body moving along our line of sight if we notice a Doppler shift in its spectral lines. A shortening of wavelength (blueshift) indicates a motion towards us, and a redshift is effected in an object moving away from us. When two bodies are gravitationally bound, they will rotate about a common centre of mass. In the case of the Earth and the Sun, our planet is so small and so distant that the centre of mass is near the centre of the Sun. As we take a long circular orbit about this centre of mass, the Sun stays in virtually the same place. However, if the planet is massive enough, and orbits close enough to the star, then the centre of mass of the combined system may be a significant distance from the centre of the star. In this case, the star will move in a noticeable orbit about this point, along with the planet.

This then is how the Swiss astronomers detected the planet orbiting 51 Pegasi. Mayor and Queloz used a spectrograph on the Haute-Provence Observatory telescope and, over a period of a year, made several separate observations of the spectral characteristics of the star. The change in wavelength clearly oscillated between a redshift and blueshift over a period of 4 days, 6 hours. Mayor and Queloz made further measurements over the following year to be sure that they had ruled out every other conceivable explanation, including a dim binary companion, instrumental error, etc. Finally they were able to show conclusively that a planet smaller than one to two times the mass of Jupiter was orbiting at 0.05 AU – less than one-seventh the Sun-Mercury separation. While 51 Pegasi is remarkably similar to our own Sun, the orbiting planet is not a good candidate for life. Given its mass, it is most likely a gas giant and, being as close as it is to its parent star, there is virtually no chance for water on its surface to be in liquid form. But the significance of this discovery cannot be understated – it represented conclusive evidence of the existence of planets outside our Solar System, or exo-solar planets as they are now known.

Since this discovery, another 180 such planets have been added to the list (and undoubtedly many more by the time you read this). There have been several notable additions:

1. star systems with multiple planets

2. planets with orbits as large as that of Jupiter
3. A solid planet only 5.5 times the mass of the Earth.

The detection of a true Earth-like planet may yet come, but this is a real challenge; the size of the Doppler shift involved may be too small to measure. Meanwhile, we will have to content ourselves with finding more large planets to get an idea of just how common they are. Other methods of detecting planets have since been developed, including measuring the dimming of a star as the planet passes in front of it (transit detection). Such research is incredibly valuable, but there was perhaps a greater desire to obtain an actual image of an exo-solar planet. And recently, just such an image was taken.

Planets are quite cold objects and generate almost no visible light of their own. As a result, we see all the planets in our Solar System entirely because of the solar light they reflect. In 2004, an international team of astronomers used an adaptive-optics-corrected VLT telescope to record the image of a brown dwarf designated 2M1207. What they discovered was a faint spot (now called 2M1207b) which seemed to be a companion to the star – shown in Plate 10. Subsequent images taken by the Hubble Space Telescope showed that the object was visible to that telescope as well. Brown dwarfs are incredibly dim stars barely able to generate any appreciable light of their own, so a fainter object nearby had to be either a distant object unrelated to the star, or perhaps a planet in orbit.

Images over the period of a year showed that the object was moving along with the star, so it was not a background object appearing to lie alongside the star. But even as a companion to the brown dwarf, its orbit is so large that it would take years to move by an observable amount. Instead, astronomers measured the size, spectrum and brightness and from the numbers obtained deduced that the only object it could be is a planet. While it is still most likely a gas giant like Jupiter, this is a great step forward. We now have the first photographic evidence of a planet orbiting another star. The next step will be to try to get an image of an Earth-like planet and then search for possible signatures indicating the presence of life. Details on the quest for this goal will be discussed in the next chapter.

Milky Way black hole

What does our galaxy look like? For many centuries this was an extremely contentious question. The idea that the stars we see lie in the shape of a flat slab was first made by Thomas Wright in 1750 – long after the debate over the Earth/Sun-centred universe had been settled once and for all. Even then, most scientists still seemed to think it logical that the Solar System held a place of importance (the centre) in the universe. The concept of our location in an 'island' galaxy which is a separate entity to 'other' galaxies could not be proved until the end of the nineteenth century when advances in astrophotography finally led to astronomers identifying the existence of individual stars in Andromeda. Once this idea began to be accepted it led to a concerted effort to find out the shape of 'our' galaxy.

In the early twentieth century, American astronomer Harlow Shapley mapped out the positions of globular clusters and noticed that they seemed to lie within a sphere centred towards the constellation Sagittarius, which he took to be the centre of the Milky Way. All of a sudden, the Earth did not even reside in a place of importance in our galaxy! Measurements since then have showed the Milky Way to be a spiral galaxy 100,000 light-years in diameter, with the Earth lying 26,000 light-years from the centre. The galaxy is around 20,000 light-years thick in the centre, falling to a mere 2,000 light-years thick at the edge[39]: by most standards, the Milky Way is a large galaxy. But while its size and shape had been fairly well determined, there was still the question of what lies at its centre.

Figure 14.4: Viewed side-on, the Milky Way galaxy would appear very much like this image of NGC 4565 (left). Viewed from above it may well look like that shown on the right. Courtesy: Sloan Digitized Sky Survey and NASA/JPL-Caltech/R. Hurt (SSC).

The centre of our galaxy is located in the direction of the constellation Sagittarius. The precise location has been inferred from the presence of a strong radio source, designated as Sagittarius A*. Observations of other galaxies suggest that it must be something with a large amount of mass in order to ensure that the furthest stars remain gravitationally bound within the galaxy. From many different types of measurements, astronomers have theorised that the Sagittarius A* radio source is actually a super-massive black hole. The black hole may never be observed directly, as the diameter of the event horizon is just 7.7 million km – only five-and-a-half times the diameter of the Sun. Instead, astrophysicists search for evidence which will ultimately lead to no other possible explanation for the observations we collect.

In October 2002, the best evidence to date was presented by a group from the Max Planck Institute for Extraterrestrial Physics (MPE) led by Rainer Schödel. These astrophysicists took a series of images of the stars nearest to the suspected black hole over a period of ten years. The observations were made with the New Technology Telescope, the Keck telescopes and finally the VLT telescopes. The result was a plot of the motion of one star (S2) which has a 15.2-year orbit about an invisible object and comes within a mere 17 light-hours from Sagittarius A*. Plate 11 shows an image of the star field along with a plot of the position of S2 relative to Sagittarius A* over a decade of observations. The image is truly a marvel when you consider that the resolution is a mere 60 milliarcseconds.

S2 comes as close as 120 AU to, and as distant as 1700 AU from, the gravitating body around which it orbits in just 15 years. Let's compare this with the orbit of Pluto, which has a similar shape. Pluto's distance to the Sun varies between 30 and 50 AU, but the planet takes around 250 Earth years to complete one single trip around the Sun. How, then, can S2 have an orbit some 40 times larger than that of Pluto, but take one-sixteenth of the time to make an orbit? Well, like objects spiralling down a drainpipe, the steeper the curve, the faster they will move. So it should come as no surprise that when we apply gravitational theory to the problem we can calculate the mass of Sagittarius A* to be 3 million solar masses. Our current understanding of physics shows that anything having this amount of mass in such a small volume of space that is not a black hole will immediately become one. Thus we can be fairly sure that this is, in fact, what lies at the centre of the Milky Way.

Hubble Ultra-Deep Field

The Hubble Space Telescope is uniquely able to observe extremely faint objects, since it does not have to contend with any interference caused by airglow. Furthermore, since it is in orbit, it can take images with incomparable exposure times. It was suggested that the HST be turned towards a particularly empty portion of the sky in order to image the most distant and oldest objects visible. With this in mind, the first long exposure image, called the Hubble Deep Field (HDF), was taken in 1995. In all, 342 separate frames were taken for a total exposure time of around 10 days over 150 orbits! A portion of the HDF image is shown in Figure 14.5.

The direction in which to point the telescope had to be chosen with great care. Firstly, there are large exclusion zones around Solar System objects due to the excessive light levels from them. Next, the direction in which the telescope is pointed should be at right angles to the plane of the Milky Way to avoid bright stars and obscuring dust. Likewise, there was a need to avoid the ecliptic and Zodiacal light as well as any portions of sky where there are nearby galaxies. Once all of this was taken into account, the final decision was made to observe in the direction of Ursa Major (the Big Dipper). The field of view was a mere 2.5 arcminutes on a side, but this is still large enough to encompass a thousand galaxies. If the observation area chosen is representative of the rest of the sky, this implies the existence of hundreds of billions of galaxies in the observable universe. But aside from the statistics, a great amount of science can be done analysing the morphology of galaxies (i.e. how they evolve over time). Over the following years, several groups took on the job of cataloguing the appearance of the galaxies in the image as a function of distance, which could be determined using ground-based spectroscopic measurements. This sort of information can help us understand how matter in the early universe was distributed and how the first galaxies formed.

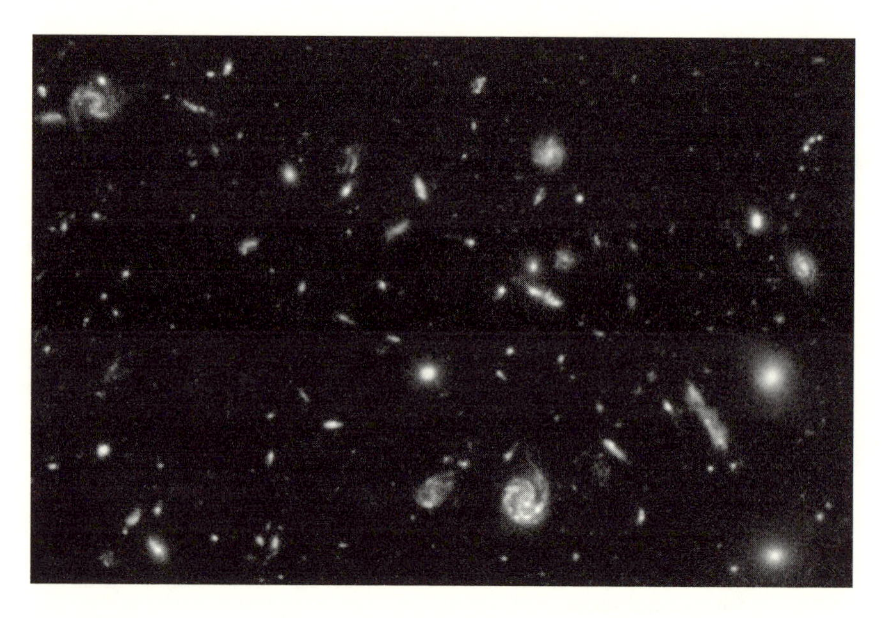

Figure 14.5: Hubble Deep Field. A greyscale reproduction of the Hubble Deep Field showing a myriad of galaxies with various sizes and shapes. Courtesy: Hubble Space Telescope (STSCI/AURA).

The success of the Hubble Deep Field led to a second image being taken in the southern celestial hemisphere – called the Hubble Deep Field South. This image, taken in 1998, encompassed a similar area of the sky with the same exposure time, but this time in the direction of the southern celestial hemisphere. Then in 2002 an improved camera was installed in the HST, which permitted higher-resolution images to be taken of fainter objects. A new portion of sky was chosen in order to capitalise on this new camera and produce an even better long-exposure snapshot. The result was the Hubble Ultra-Deep Field released in 2004 and shown in Plate 12.

This time the telescope was pointed at the constellation Fornax and the image encompasses a 3 arcmin field of view in the visible. The exposure time was 1 million seconds (11.5 days) long. It required over 400 orbits, and needed to be this large as the faintest galaxies pictured as little as 1 photon per minute. In this image we can see objects as faint as 30th magnitude – one trillion times fainter than the star Vega. Some of them could be as distant as a redshift of 10–12, which means that we are looking at light that dates back to when the universe was as young as 800 million years old. There may be as many as 10,000 galaxies visible in this image that will provide years of enjoyable analysis for many astrophysicists. It is perhaps daunting to think of the fact that in each of these galaxies there are (now, anyway), some 10 billion stars. That makes over 100 trillion star systems. It is hard to doubt that around one of these stars there are other life forms which are taking their first steps in looking at the universe around them. Perhaps they are looking at a similar image of our region of the universe taken back across the same vast reaches of space, pondering if there is life over here.

Hoag's Object

Perhaps the most photogenic of heavenly bodies (other than supermodels) are planetary nebulae. These clouds of colourful gas form towards the end of the lifetimes of stars with low or medium masses, like our Sun. After burning most of their hydrogen, the stellar core begins to contract and heat up to burn helium. This causes the outer layers of the star to expand and cool and the star becomes a red giant. Over time, as the core temperature increases further, the outer layers of the star start to get blown out into space. This process continues for years, eventually producing a diffuse cloud of material around the star, up to a light-year in diameter. Meanwhile, the central star continues to contract and heat up, eventually getting to a stage where most of the emitted light is in the ultraviolet part of the spectrum. As this happens, the starlight heats up the surrounding cloud and causes it to glow in the colourful display that we call a planetary nebula. This phase of the process lasts about 10,000 years – a mere blink in the many-billion-year lifetime of the star. Eventually the star will lose the ability to burn fuel as it fades into a smoldering white dwarf.

There are about 1,000 planetary nebulae that have been catalogued, with as many as 10,000 suspected to exist in our galaxy alone. They are usually quite spectacular in their range of colours and shapes. As a result, they have lent themselves to imaginative monikers such as Eskimo, Hourglass, Ant, Egg, Owl, Dumbbell, Spirograph and Cat's Eye. It was perhaps no surprise then, when in 1950 an American astronomer, Art Hoag, announced the discovery of an unusual object in the constellation Serpens. It had the look of a planetary nebula with a blue ring surrounding a bright yellow central 'star'. But initial spectral analysis by Hoag suggested that Hoag's Object, as it has become known, consisted of stars – which would make it a galaxy in its own right and not a nebula lying in our galaxy. This was backed up by further measurements made in 1974 that set the distance to Hoag's Object at around 600 million light-years – and also meant that it is about the same size as the Milky Way. But even more confusing to astronomers was its unusual shape.

Galaxies of many different shapes and sizes are known, but they generally fit into a few well-defined types allowing for easy cataloguing: spiral, barred spiral, elliptical and irregular. For the most part, these shapes closely resemble the patterns foam can form on the surface of a stirred cup of coffee. It is believed that galaxies formed from small density fluctuations in the early universe that served as seeds, slowly collapsing under gravity and accreting more and more material into a central core. This distribution of matter results in a gravitational gradient that slowly decreases towards the edge. Since matter is arranged by gravitational forces, the distribution of stars will more or less follow the same pattern. In your coffee, the gravitational gradient is approximated by the depression in the surface of the rotating liquid, and the foam is like the visible galactic matter. Now look at the image of Hoag's Object as shown in Plate 13 and ask yourself: have you ever seen foam in your coffee turn into something like this?

It appears that the galaxy consists of a ring of predominantly blue stars surrounding a dense core of yellow stars. But why the stars continue to orbit in a ring without coalescing into a uniform disk or spiral is unknown. Several theories have been proposed but none have yet been conclusively proven. Hoag's Object thus represents just what telescopes can offer us. Not only can they display unimaginable heavenly beauty, but they also highlight the fact that there are many mysteries out there to be answered. As our telescopes bring us more detailed images of the cosmos, we increase our understanding of the heavens and the physics which shapes it. But even as we start to think we know it all, more aspects are revealed which replace our answers with yet more questions. As a scientist, I find this delightfully frustrating.

Chapter 15

Future telescopes

To get an idea of the future of telescope design, you only need to look in your history books. The last 400 years has seen a progression of steady increases in mirror diameter, with a doubling of aperture every 50 years or so. This trend is far from over, and, in fact, it appears that we will do even better in the future. Meanwhile, as technology improves, so too does our ability to capture and analyse light. In this chapter we will look at the next generation of telescopes currently on the drawing board. Some are designed for general imaging while others address a specific question we want answered. All of them are bound to make the headlines.

Wide-field wonders

Wide-field telescopes are very useful for creating catalogues of celestial objects, but this is more than an astronomer's version of stamp collecting. To begin with, there is the measuring of stellar positions, called astrometry, which is vital for improved pointing and positioning of satellites. Much as sailors navigated by the stars in ancient times, so do space probes require stars as a method of knowing which way they are pointing. The Hubble Space telescope uses a star catalogue of some 19 million objects for this purpose.

Wide-field telescopes are also necessary for gathering statistics on large numbers of objects in a reasonable amount of time. For instance, at the Anglo-Australian Observatory in Australia there is a project called 2dF. This instrument is designed to gather light from hundreds of galaxies over the two-degree field of view (hence the name) and measure their redshifts. Converting redshift into distance makes it possible to build up a three-dimensional map of the locations of these galaxies. The result is shown in Figure 15.1, where the distribution of over 220,000 galaxies is displayed. Before the first maps like these were made, it was thought that all the galaxies in the universe were distributed evenly, but instead we can clearly see that there are large 'voids' and 'filaments' where galaxies are packed closely together or are spaced far apart. The discovery of this large-scale structure has had a significant effect on theories of the formation of the universe and Grand Unified Theories aimed at merging quantum physics with General Relativity.

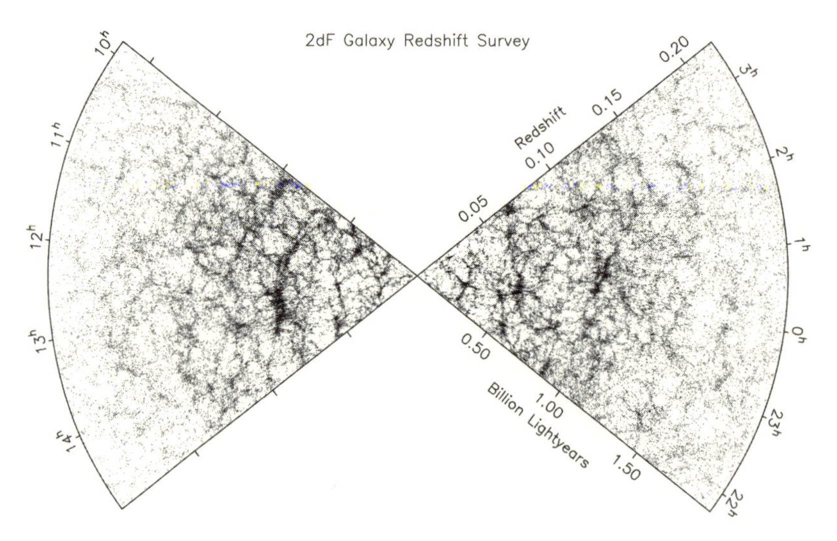

Figure 15.1: A slice of the universe showing the positions of thousands of galaxies (each one represented by a single point). Our location (the Milky Way galaxy) is in the centre of the image. Courtesy: 2dF Galaxy Redshift Survey Team.

As well as being useful for catalogues and sky surveys, wide-field telescopes can be used to detect changes in the appearance of objects in the sky from one point in time to another. One such project, being funded by the United States Air Force, is called the Panoramic Survey Telescope and Rapid Response System (Pan-STARRS, for short). This observatory is to consist of four 1.8-m diameter, wide-field telescopes located next to one another. One reason for having four telescopes instead of one is redundancy. Making large-scale CCD chips for astronomy is extremely difficult. Even when the greatest care is taken, there can often be many 'dead' pixels which do not respond to light or fail to transmit the data properly. This is not a problem with conventional telescopes, as these pixels can often be ignored, but this is not acceptable for the work Pan-STARRS is planning to do. By having four telescopes each taking the same image, the effect of dead pixels can be eliminated by making a comparison of the four images collected. For example, if the same pixel from each of the four telescopes is read out to be '1250, 1250, 0, 1250', then we can safely say that that there is something wrong with the pixel in the third telescope and it can be ignored. The four telescopes also serve another purpose: the same observation can be made through different filters simultaneously.

Pan-STARRS is designed to take 1000 images of a different 10-square-degree patch of sky every night and to revisit each of these patches every week or so. If a bright spot has shifted position with respect to the background stars during this time, this would indicate a moving object. The Air Force is supporting this telescope for the optical sensing of orbital objects. Standard radar tracking is limited to detecting objects 10 cm across or larger and there are some 10,000 objects of this size or larger, including around 700 satellites. Using optical methods, smaller objects can be identified and catalogued. This includes a multitude of bits and pieces that have been discarded during past missions or simply fallen/broken off larger objects. Space junk is a significant problem for both equipment and personnel as even a fleck of paint at orbital speeds can have as much kinetic energy as a bullet. In fact, there have been three confirmed collisions in orbit since 1991 – a rate much higher than expected. With most satellites costing hundreds of millions of dollars, an improved understanding of the orbital environment is worth some investment. Quite apart from the military applications though, Pan-STARRS will detect other objects that move above us. These could be asteroids, comets or perhaps pieces of rubble orbiting around the outer fringes of our Solar System. Astronomers can calculate the orbit of a body using knowledge of the star positions and the time between exposures. From this we can find out if an encounter with the Earth is likely. To me, this is probably the most significant aspect of the Pan-STARRS effort.

If you are in the Eastern hemisphere on the night of Friday 13 April 2029, you will witness one of the scariest events of your life. Looking in the direction of the constellation Cancer, you may see a fairly bright spot (magnitude 3.3) moving across the sky over the course of an hour or so. The 300-m diameter asteroid with the catchy name of 99942 Apophis (formerly the equally catchy 2004 MN4) is scheduled to pass within 30,000 km of the Earth's surface. Sounds like a lot, but this is actually inside the orbit of geosynchronous satellites. Calculations have made it all but certain that no collision with the Earth is imminent, but the danger won't end there. The near miss will dramatically change the orbit of the asteroid so that it may come back to hit us on 13 April, seven years later. If you are superstitious, it may ease your mind to know that this will be a Sunday. Current estimates put the chances of collision at only 1 in 5000, but given that the resultant energy release would be 15 times greater than the largest nuclear weapon ever detonated, it is still a little disconcerting.

Since the middle of the nineteenth century, palaeontologists have documented signs of periodic mass extinctions occurring throughout the Earth's history. Such extinctions are very dramatic – often killing off as much as 95% of the world's fauna in a matter of years or less. The last one, which occurred about 65 million years ago, was responsible for wiping out the dinosaurs and ended the Cretaceous period (it marked the so-called K-T boundary). In 1980 scientists from the University of California, Berkeley announced that they had discovered that excavations of rock at the K-T boundary contained abnormal amounts of iridium in them – some 30 times more than expected. Iridium is a fairly rare element on Earth, and in fact one of the few places where we know we can find such high concentrations of it is in meteorites. The palaeontologists used this as evidence for their theory about what killed the dinosaurs. It begins with a huge meteorite or comet that slammed into the Earth, spewing out an enormous amount of dust into the atmosphere. This dust blocked out most of the light from the Sun to create an extreme artificial winter which lasted for years. The cold-blooded dinosaurs could not adapt to this dramatic climate change, while the smaller, warm-blooded mammals were barely able to survive and go on to become the progenitors of most of today's species, including *Homo sapiens*.

Further investigations showed that thicker layers of iridium were to be found in Central America – suggesting that the impact took place somewhere in that region. Then, in 1990, a group led by Alan Hildebrand discovered what seemed to be evidence of a crater in the Yucatan Peninsula to the east of the Gulf of Mexico. The 180-km diameter Chicxulub crater (pronounced *Chick-shoe-lube*) had all the hallmarks of the smoking gun (or perhaps gaping bullet wound would be a more appropriate metaphor). Modelling has since shown that the meteor was probably around 10 km in diameter and was travelling at around 50,000 km/hr. The energy released in the collision was equivalent to 300 *billion* times that of the Hiroshima atomic bomb. Material was ejected hundreds of kilometres 'downstream' and over 50 km into the atmosphere. Not surprisingly, the impact created Magnitude 9 earthquakes that were felt all around the globe. No doubt a few dinos knocked off their feet in Central America were more than a little perplexed, but it was merely the signal of worse to come.

Bits of rock strike the Earth all the time (up to 100 tonnes' worth per year), mostly a few centimetres or less in diameter. Scientists have calculated that an asteroid or comet in excess of 5 km in diameter would be sufficient to cause a similar extinction of human life. The current thinking is that a bolide this size should strike the Earth about once every 10 million years or so. While this may not seem too much to worry about, the fact is that this means that the likelihood of dying from an asteroid strike is roughly the same as that of dying in a plane crash, so it is not unreasonable that some money should be spent on addressing the issue. The first step is to try to locate and characterise all the Earth-crossing asteroids in our Solar System – a daunting project, given the huge amount of space to be covered combined with their high speed, small size and low reflectivity.

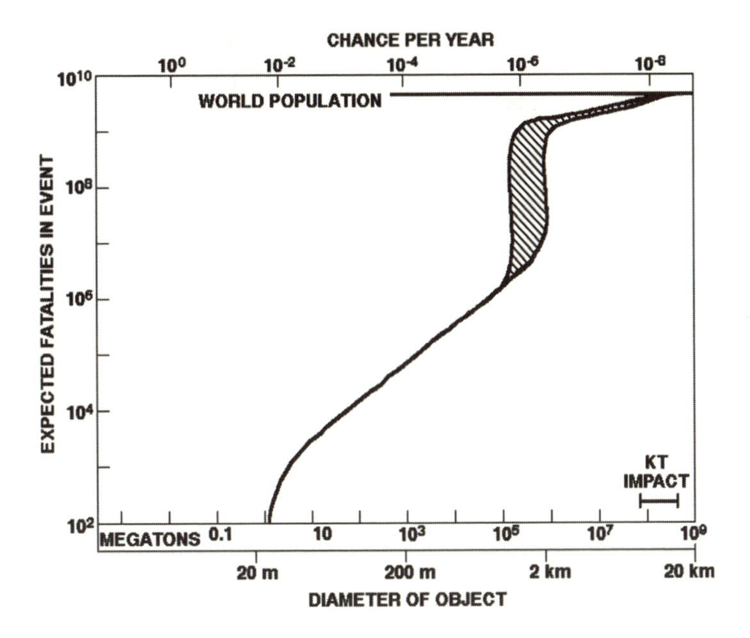

Figure 15.2: A graph showing the likelihood of certain asteroid collisions, along with the size of the impactor and probable number of fatalities. The grey area is bounded by the upper and lower limits. Courtesy: Clark Chapman (SWRI).

Using the huge amounts of data being collected by Pan-STARRS, a complete catalogue of 99% of all potentially hazardous objects with diameters larger than 1 km should be completed by 2020. In fact, this telescope will discover more asteroids in the first month of operation than the 3000 currently known. The significance of this effort cannot be over-estimated – it represents a big step towards greater understanding of the potential extinction events governing our existence. However, Pan-STARRS will be doing more than discovering new asteroids and comets. Anything that changes in brightness from one exposure to another will be catalogued. For astronomers, this will mean a huge amount of data on variable objects of all kinds. But for this work, a better telescope is on the drawing boards – called the Large Synoptic Survey Telescope (LSST).

LSST will be an 8.4-m diameter telescope capable of imaging a whopping 10-square-degree field of view. The combination of its primary collecting area and field of view (known as entendue) is more than ten times that of any existing telescope and five times that of Pan-STARRS. The telescope will be able to re-image the same piece of sky every three to four days down to a magnitude of 24, making it possible to develop a huge image database of faint objects. The LSST design consists of a three-mirror system feeding into three lenses designed to further reduce off-axis aberrations over the large image plane. At the heart of the telescope lies a mosaic of CCDs with a total of 3 gigapixels – some five hundred times more than in a standard digital camera. Needless to say, with the LSST imaging a portion of the sky every ten seconds, the amount of data generated will be huge. One image will be 36 gigabytes in size, and in a single night, the amount of data collected will be around 30 terrabytes – enough to fill over 6,500 DVDs. In the course of a year, the amount of data will be 200 times this, so one of the big challenges will be to develop new techniques for handling and storing such huge quantities of data.

So why build such an instrument in the first place? As we have already seen, wide-field telescopes are useful for imaging large numbers of objects at a time. This can be applied to simple catalogues or used for observing changes in the night sky from one night to another. Like Pan-STARRS, the LSST will detect and characterise new asteroids and comets. In fact, if Clyde Tombaugh had had the use of a telescope like LSST, there is every chance that Pluto would have been discovered within a week. But with such a large collecting area, the LSST will be able to see much fainter objects – over a billion of them over the entire sky. This makes it ideal for looking for variations in the brightness of faint objects, and will lead to the discovery of thousands of variable stars, supernovae and lensing events. Variable stars and supernovae can be used as standard candles to improve our knowledge of the size, structure and morphology of the universe. Lensing events, on the other hand, have a completely different use.

General relativity predicts that all masses create a curvature in space-time which will in turn affect the motion of other masses and even light itself. Einstein realised that, given the right circumstances, this might allow large masses to act like lenses. It works like this: if light from an extremely distant object passes close to a very large intervening mass (e.g. a galaxy) the light could be bent, much as refracting light is bent in a lens. However, unlike a glass lens, the gravitational curvature is not conveniently shaped to give an image of any significant clarity. More often than not, the lensed image we see is distorted beyond recognition – it appears either as multiple blurs, thinly stretched blobs or perhaps a ring. While Einstein came up with the concept of a gravitational lens, he thought it would be extremely unlikely that the arrangement of matter in the universe would conspire to make it possible to see the effect. He was mistaken, but it did take a while to find one.

The first image of a gravitationally lensed object (a quasar) was made in 1979. Since then many more examples have been found, mostly involving extremely distant sources and lensing by massive galaxies. We can use lensing events in many ways. To begin with, they can provide us with a method of 'seeing' otherwise invisible phenomena in the universe. From measurements made by many different observatories, we now know that the luminous universe constitutes a mere 4% of everything 'out there'; the rest is invisible dark matter and energy. While we can't see the majority of the cosmos, we can get an idea of its makeup and distribution by observing the effect it has on the material we can see. In order to do so, we need to observe extremely distant (dim) objects in great numbers. This requires a large-diameter telescope which can also image massive areas of the sky. While this is the major reason for building the LSST, it will serve another purpose.

In the early 1990s a group of Australian and American astronomers formed a joint team to try to detect microlensing events, or lensing from smaller and closer sources. The MACHO group (a catchy acronym for Massive Compact Halo Objects), was specifically looking for the chance alignment of a dim star in our galaxy with a star in a nearby galaxy. The expectation was that, as the nearby star wandered in front of the distant star, we would see the distant star slowly brighten then dim as the nearby star gravitationally focuses the light to the Earth. In 1997, the MACHO group announced the results of its first photometric survey of thousands of stars in the Large Magellanic Cloud with the 1.25-m telescope on Mt Stromlo. It had found 45 such candidate events. An example of such an event is shown in Figure 15.3.

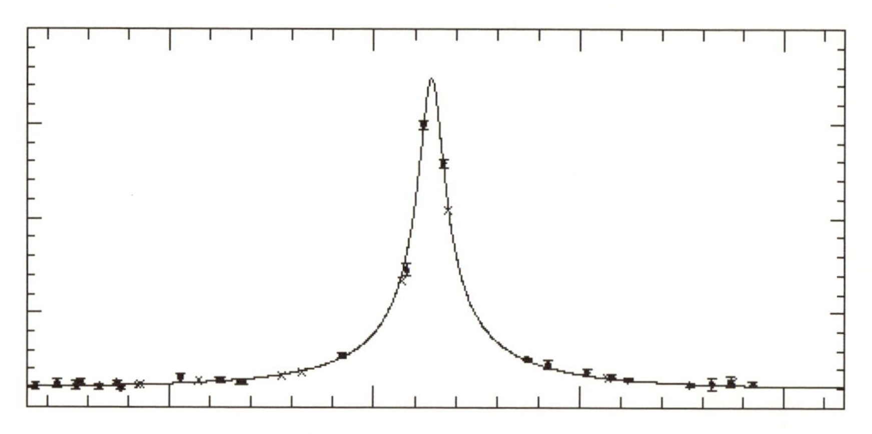

Figure 15.3: This plot over a period of about a month shows the brightening of a distant star caused by the gravitational lensing of a nearby star. Courtesy: MACHO group.

Since this discovery, many more microlensing events have been recorded by this and other groups. Perhaps the most intriguing discovery came when the MACHO group announced the discovery of a light curve with a multiple peak, which suggested the presence of a Jupiter-sized planet orbiting a binary star system. Since then, several more candidate planets have been identified in this manner, including the smallest one found to date – a mere 5.5 times the mass of the Earth. The success of the MACHO project is one of the big drivers behind building the LSST. The objective is to observe millions of sources on a daily basis and detect thousands of these microlensing events. In turn, the information gathered will help us find out more about the populations of low-mass stars as well as planetary systems. Ultimately the goal of all such planetary studies is to find a true Earth-like planet in orbit around another star.

Another pale blue dot

In the previous chapter we discussed the discovery of planets around other stars using Doppler techniques. Over 180 have been catalogued, with the smallest being less than the mass of Neptune. In 2000, the first planet was observed dimming the light from a parent star by passing in front of it. Since then, this transit method of detecting planets has turned up several more candidates. With this in mind, NASA is planning to launch the Kepler Telescope in 2007. The 1-m diameter telescope is designed to continuously look in a single direction (near Vega), keeping constant watch over the light intensity from around 100,000 stars within a patch of sky as large as your fist with your arm outstretched. It is hoped that over the course of the mission up to 50 Earth-like planets will be detected transiting their parent stars. While Kepler and the many ground-based telescopes will undoubtedly gather evidence of a menagerie of Earth-like exo-solar planets within the next decade, it appears unlikely that we will be able to image such pale blue dots with existing instruments. The reason, as we saw in the previous chapter, is due to scattering in the atmosphere which causes the light from the parent star to swamp the light from the planet. Even with a space telescope there are still many issues to be resolved.

Let's begin with a simple calculation of the size of a telescope required to resolve Earth from the Sun (1 AU) at a distance of, say, 30 light-years. Using our resolution equation once again, we can calculate the size of the telescope needed for the job to be just over a metre in diameter. This would suggest that the Hubble Space Telescope (with a diameter of 2.4 m) should be able to produce images of such a planet quite easily. The problem is that the resolution criterion only works for two point sources of the same brightness. Unfortunately, a terrestrial planet will be about one billionth as bright as the parent star, so the Airy spot from the star will overwhelm the light from the planet. To put this in perspective, consider Figure 15.4 showing the ideal Airy spot of a star. Even if the Airy spot of a planet is located at the position of the point of zero intensity (indicated by the arrow), it would be less than one millionth the thickness of the line – and be completely invisible. In order to have more chance of seeing the planet, we have to further separate the planet from the star, which requires using a larger telescope. Either that or we need to find a way of blocking out the light from the star. But of course, as always, it still isn't quite as simple as this.

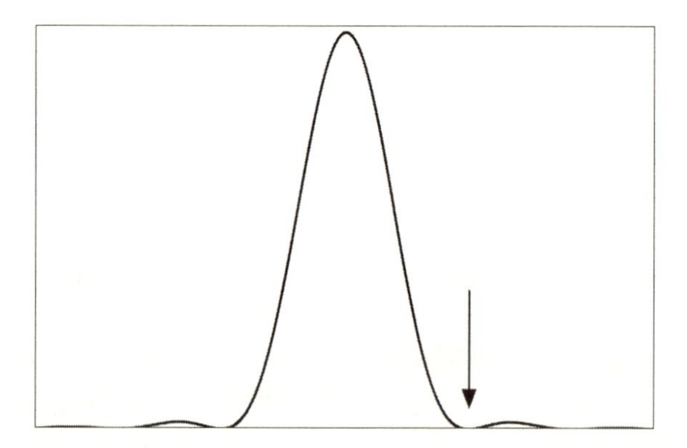

Figure 15.4: The Airy spot of the star is many times brighter than the Airy spot of the planet located at the position indicated by the arrow.

The Airy spot doesn't quite tell the whole story when we are trying to view very dim objects, as in real optical systems the mirror roughness starts to become important. We have already determined that a mirror ground and polished to a diffraction-limited surface can have a bump no larger than $\lambda/8$ and it is a fairly straightforward procedure to fabricate a telescope mirror to this quality. However, while this sort of mirror will produce a nice Airy spot, surface features at much smaller scales will scatter small amounts of light to other places on the image plane. Think of it as being similar to seeing a reflection of the Sun on a lake. A smooth, quiet surface may reflect a very clear image of the Sun, but the slightest ripple anywhere else on the lake may also send a glint of light into your eye. Likewise, a small amount of surface roughness, scratches or dirt on a high-quality mirror may divert a tiny fraction of light off in the wrong direction. In planet detection, we need to keep this stray light well below the level of one photon in a billion, which means the surfaces have to be smooth down below the nanometer level. We must also make sure that all dust and coating defects are eliminated as they too will scatter light.

The Terrestrial Planet Finder (TPF) actually consists of two separate missions, taking widely different approaches to the same problem. The first telescope is TPF-C, in which the 'C' part stands for 'coronagraph'. A coronagraph is a telescope modified to block out light from a star. By placing a circular disk over the image of the star, we can remove the starlight from the image and, we hope, see the planet lying next to it. The second planet finder mission is an interferometer, hence the name TPF-I. As we have seen, interferometers are capable of ultra-high resolution imaging over a small field of view. What's more, with nulling interferometry, it is possible to cancel out the light from a star, which is a beneficial property for imaging planets.

A coronagraph gets it name from the original application, which was to block out light from the images of the Sun's disk so that the weaker light from the surrounding corona could be seen. Quite serendipitously, the Moon is often just the right size to perform this same function during a solar eclipse, and this was how the corona was first discovered. Today, the Solar Heliospheric Observatory (SOHO) spacecraft uses a coronagraph to permit us to take regular photos of the Sun's outer 'atmosphere' like that shown in Figure 15.5. Its ability to see dim objects near the Sun has also made this space probe a prolific comet hunter – it has been used to find nearly half of all those known. Most comets go undetected as they may not brighten until they pass close to the Sun, where they are simply too dim to be seen in the bright glare. The SOHO coronagraph makes them visible again, and in August 2005 it racked up its 1000th new one. The TPF coronagraph telescope will work on the same principle – not for detecting comets, but for finding exo-solar planets.

Figure 15.5: An image of the Sun taken using the SOHO coronagraph. The Sun's streaming corona, a comet (bottom, left of centre) and even background stars (towards the centre of the Milky Way) can be seen with the solar disk obscured by the stop. Courtesy: NASA/ESA.

TPF-C will have a somewhat unusual telescope design, as shown in Figure 15.6. The first thing to notice is the off-axis mirror which is designed to direct the focused light off to one side, out of line with the incoming light. In the case of most conventional telescopes, we have the problem of how to hold the secondary optics in front of the primary. The usual solution is to use a 'spider': a few metal bars which come in from the side of the telescope framework. But as we saw in Figure 4.1, any barrier in front of a wavefront will bend (diffract) a small amount of light off in all directions. In the case of a four-pointed spider, this light will make every star look like a plus-sign, which are often seen in astronomical photos such as those in Plate 14. In fact, in this example, there are additional haloes around each star caused by multiple reflections of the starlight within the Schmidt plate of the classical Schmidt telescope. Diffractive and reflective effects of this sort are not acceptable in planet-finding telescopes as they would obscure a planet. For the Terrestrial Planet Finder, the solution was to reflect the light out of the way of the incoming beam. You don't normally see telescopes designed like this, as there is generally no need to avoid the tiny amounts of diffraction which turn the stars into spikes. Besides, when a telescope is on the ground there will be so much scattered light and blurring from the atmosphere that there would be no point.

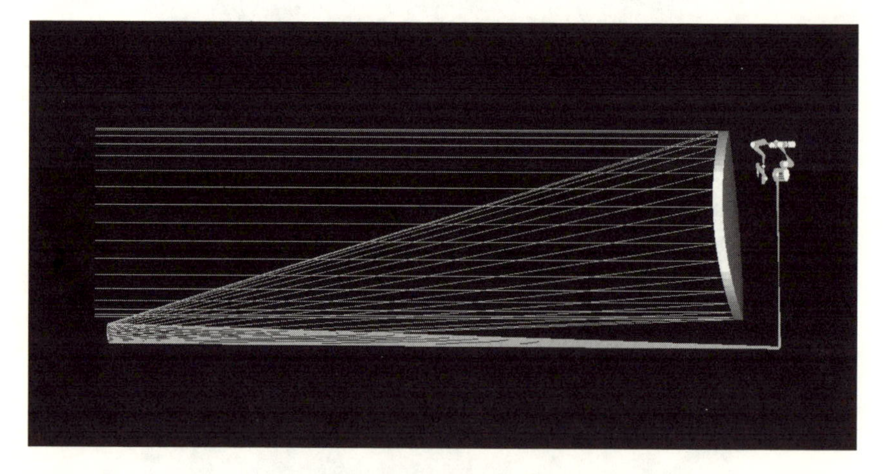

Figure 15.6: The optical design for TPF-C. The light (entering from the left) is reflected off the primary to an off-axis secondary and then to smaller optics behind the primary.

The TPF-C primary is a hyperbolic mirror with an elliptical cross-section; 8 m in one dimension, and 3.5 m in the other. The main reason for choosing such a strange shape is that it was the largest single mirror shape which could fit down the length of a rocket. For this experiment, there is no option for using a segmented mirror as the edges of the mirror segments would diffract light in much the same way as a spider. As large as this mirror is, though, the surface still has to be good to about $\lambda/3000$ or better. To put this number into perspective, if the mirror were expanded out to the size of Australia, the biggest permissible bump would still only be about the thickness of a sheet of paper.

The primary reflects light onto a secondary that directs the light behind the primary and deep within the spacecraft to where an image of the distant star is produced. It is in this later stage where the opaque blockage is located. This disk, or stop, is specifically designed to cut out the starlight while allowing the light from the planet to continue to the final image plane where the CCD lies. In principle, the telescope is relatively simple, but there are more complexities (such as other types of stops) which may be added in the final design. The engineering difficulties lie in the details of how to design the optics and launch the instrument so that it performs to specifications. And this is the easier mission of the two.

The second planet-finder mission, TPF-I, is even more ambitious. This experiment will use many of the interferometric techniques mentioned in Chapter 7, but since these are difficult to manage even on the ground under controlled conditions, they are an incredible engineering challenge in space. TPF-I will work as a nulling interferometer in order to cancel out the light from the parent star. Meanwhile, light from the orbiting planet lying off to one side will not be cancelled out. In this interferometer there are to be four telescopes, each 4 m in diameter, spread out in a line up to 100 m in length. Hard to believe given that the largest telescope currently in space is a mere 2.4 m in diameter. Oh, and did I mention that the telescopes are not going to be tied together? Instead, they will fly in formation – directing their light into a fifth spacecraft, which has the job of recombining the beams. An artist's conception of this interferometer is shown in Figure 15.7.

Even though TPF-I operates at infrared wavelengths some ten times larger than that of visible light, it is an extremely ambitious project and represents the very limit of what we think we can manage with current technology. In fact, the project may require the prior launch of a separate test-bed spacecraft in order to better understand the problems involved in formation flying. After all, for TPF-I to work, the path matching of the light combined from each telescope must be accomplished to 1.5 nm over days at a time. Since there is no friction in space, it is difficult to move objects to a certain position and keep them there. The spacecraft will experience the continual push and pull of gravitational forces and solar radiation pressure but will have to be kept perfectly still for the experiment to work. All in all, it promises to be a nightmare for the engineers involved.

Figure 15.7: An artist's impression of TPF-I. The terrestrial planet-finder interferometer will require four 4-m telescopes flying in formation with a fifth beam-combining spacecraft. The thin lines joining the telescopes are the light beams. Courtesy: NASA/JPL-Caltech.

It is hoped that with the TPF missions, it will be possible to cancel out starlight by a factor of almost 100 million which may make it possible to image Earth-like planets in nearby planetary systems. Of course, if you are expecting a large image showing oceans and continents, you will be a little disappointed. TPF will merely provide a few pixels of light as proof of the existence of a planet. While images of these new planets are bound to capture the public imagination, scientists will gain very little information from them. As yet it is unknown exactly what instruments will be incorporated into the telescopes, but it is likely that a spectrometer will be necessary. If a spectrum of the light can be gathered, then we can begin to check for signs of life, for example with evidence for the presence of water and methane. This organic molecule is mostly generated in large quantities by living matter but is rapidly broken down by sunlight. If methane is detected in a planetary atmosphere it could possibly signal the existence of a life-form which is constantly replenishing the supply into the atmosphere.

At one stage the timetable had TPF-C seeing first light in 2014 and TPF-I coming online around 2020. Recently though, they were put on indefinite hold by NASA as money is diverted to the manned space programme, so the future of the telescopes is unknown. TPF are high-cost missions, but their scientific and cultural impact could be priceless. To see why, we only need consider the first images of the Earth taken by spacecraft in the 1960s. They showed that we all inhabit a lone orb in the vast isolation of cold space. Although we had all known this for a while, actually seeing such an image had a profound sociological effect on humanity. It was the beginning of our understanding of just how precious and rare life is, and of our need to protect it. There can be no doubt that the first images of another planet like ours will cause a similar change in how we view our position in the universe at large.

The big boys

Telescopes can highlight the most remarkable insecurities in their owners, and depending on how they decide to categorise it, a number of groups claim to have the 'largest telescope in the world'. Some will use the 'largest mirror' but even this can be split between the segmented mirrors and monolithic mirrors. You could perhaps consider the largest collecting area, but then there are telescopes which use many large separate mirrors combined and others which have a single big mirror. By carefully choosing their metric, up to five or six observatories can lay claim to be the title of 'world's biggest' in one way or another. It may sound quite childish, but the pride and considerable investments of entire continents can be at stake. It is little wonder, then, that the next round of telescopes are being designed with an even greater emphasis on one-upmanship. The optical design engineers will often take the politically correct approach and assure you that this is not the case, but don't you believe it! The two major players in the next generation of ground-based observatories will be the United States and Europe. The United States is planning two telescopes in the 20-m to 30-m range: the GMT and the TMT. Meanwhile, a group from the European Southern Observatory is working on a 100-m behemoth called OWL.

The Giant Magellan Telescope (GMT) is the brainchild of a group of American universities which were also responsible for the 6.5-m Multiple Mirror Telescope and Magellan Telescopes as well as the 8.4-m Large Binocular Telescope. Together with the Smithsonian and Carnegie Observatories, they have gathered enough support to build a giant telescope of a remarkable and unique design. Rather than taking the single-mirror approach or segmented design, the group has decided to take the middle road. The GMT primary consists of seven 8.4-m diameter mirrors all on the same mount, acting as a single mirror. The final primary (shown in Figure 15.8), will have a 25-m maximum diameter, and the equivalent collecting area of a 22-m diameter mirror.

Figure 15.8: The Giant Magellan Telescope (GMT). This telescope consists of seven large mirrors on a single mount. A person standing on the base gives an idea of scale. Courtesy: Carnegie Institution of Washington.

The size of the mirror 'segment' was chosen as this is the largest single mirror which can be fabricated by the Steward Observatory Mirror Lab at the University of Arizona. The clever technique used at this facility is to have the entire oven spinning while the glass is melted and then cooled. The spinning molten glass will take on a slight bowl-shaped figure even before figuring and polishing. This saves tremendous amounts of time and money in making the mirrors. Two mirrors were made in this way for the Large Binocular Telescope. While not the lightest option, this scheme does present a low-risk (and low-cost) approach for all the next-generation ground-based telescopes. According to the current schedule, the GMT should see first light from a Chilean site sometime around 2016.

The Thirty Meter Telescope (TMT) is a telescope being proposed by another North American consortium. Jerry Nelson, the principal architect behind the Keck telescopes, is capitalising on that success to create an even larger segmented instrument, with nearly ten times the light-collecting area. An artist's conception of the telescope is shown in Figure 15.9. The first thing to notice is the similarity between this telescope and the two Kecks. The reason for this is that much of the risk behind the construction is removed by sticking to known technologies. The current design has 738 hexagonal segments, each 1.2 m in size. Like the GMT, the overall design is a Gregorian which strays from the current preferred Richey-Chrétien design. While the preliminary design has been formed, much work remains to be done before this project sees first light by the 2015 schedule. Meanwhile, however, the Europeans are looking at building a telescope with some ten times the collecting area of the TMT.

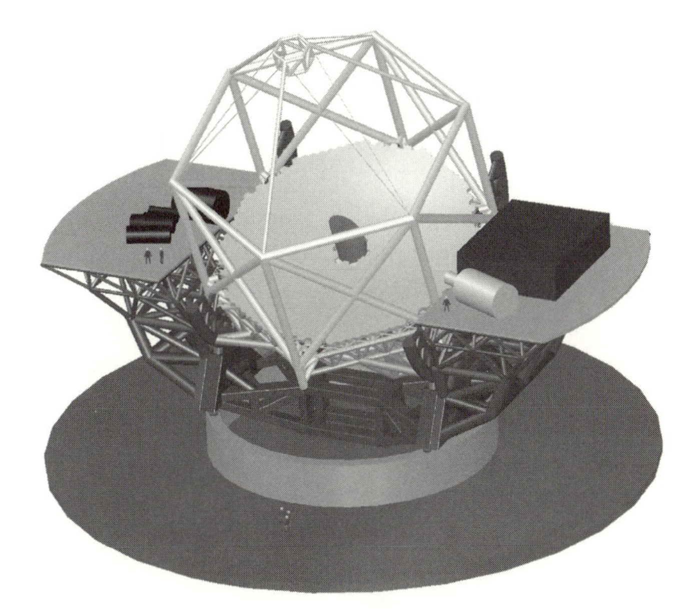

Figure 15.9: Artist's impression of the proposed 30-m TMT telescope. Three people, standing at the base and on the Nasmyth platforms, give an idea of the scale. Courtesy: TMT Project, NOAO/AURA/NSF.

In 1998, at a conference in Hawaii, I had the privilege of being in the audience during the first presentation for the concept of a 100-m telescope. In order to understand how this talk was received, it is necessary to view it in terms of the existing facilities. At that time, the second 10-m Keck telescope had been operating for only two years and only one of the four 8.2-m VLT telescopes had been commissioned. Naturally, then, talk of something ten times bigger was greeted with more than a little skepticism and some ridicule. The numbers presented were simply too incredible to take seriously; for instance, the collecting area would be larger than every single telescope in history, *combined*. Even though the speaker was careful to demonstrate that there were no foreseeable engineering reasons as to why this could not be achieved, there were more than a few giggles audible during the presentation. Perhaps this was due in part to the slightly tongue-in-cheek acronym given to the project: the Overwhelmingly Large Telescope, or OWL for short. Whatever the reasons, such a project was simply beyond the imagination of most people and was widely dismissed as something of a flight of fancy. However, the next major presentation two years later was taken much more seriously. Hard numbers, a clear project timeline and cost estimate had emerged – and the astronomical community began to warm to the idea. By the time of the next conference two years later, the project had ceased to be a question of if, but had become a question of when.

The OWL telescope is a truly audacious concept, but through some very clever design approaches, the costs have been kept quite reasonable. Consider for a moment the optical components. The current design has the primary consisting of 3048 hexagonal segments each 1.6 m in diameter. Making a single parabolic primary mirror segment of this size would typically take around six months, as each parabolic mirror segment is cast, figured, polished, tested and coated to diffraction-limited surface accuracy. At first glance, then, it would seem that the fabrication of all the segments would take around 1500 years or so. The first partial solution to this problem was to use economies of scale in the production process. For the fabrication of these mirrors, one or more entire factories will need to be built, devoted solely to producing multiple mirrors in parallel. With a production timetable of roughly one mirror per day, it should be possible to make all the mirrors in a mere ten years. Still, even with a massive production line devoted to these mirrors, this output would be impossible without a further clever design feature.

Most large telescopes are made with either a hyperbolic or a parabolic primary mirror. Since both of these shapes require a change in curvature as you go from the centre to the edge, segments of such a mirror are all different. Each mirror has to be rigorously tested to ensure it has a perfect overall shape. This in turn means that many separate testing set-ups would be required. For OWL, the solution was to design a telescope with a spherical primary made up of spherical segments – an approach used by HET and SALT. A spherical mirror has the same curvature everywhere, so every segment can be made identical – something which greatly reduces fabrication time and costs. No doubt you are quick to spot the problem here: spherical mirrors are not used in telescopes as they introduce spherical aberration. This is indeed the case, so the OWL team designed a telescope with six major mirrors, three of which have to be given some odd shapes in order to remove this aberration before the final image is formed. Although this means more mirrors, they are much smaller than the primary (but still in the 8-m range) so the cost savings are still huge.

The optical design represents just one of the many engineering problems; there are also interesting mechanical engineering issues involved in the structures. Normally, it is preferable to have a telescope completely enclosed to reduce the effects of wind and to protect it during the day. In this case, the requirement of a rotating dome some 150 m in diameter, with a 100-m opening for the light to enter would be prohibitively expensive. The solution is to have a roll-away enclosure which is moved on tracks to expose the telescope to the open air. This saves money, but puts more constraints on the design of the structure. The telescope truss must be stiff enough to resist bending under gravitational loads and winds, while being light enough to still allow precise tracking. Using advanced design analysis, the mass has been reduced to a manageable amount and looks something like that shown in Plate 15.

These are just a few of the complex issues to be resolved, but the OWL management believes that all the foreseeable problems can be solved using current technology. The one caveat may be the adaptive optics – there is simply no way of knowing whether an adaptive optics system could be constructed to make this telescope operate anywhere near the diffraction limit. If it could, the leap in image quality would be something like that shown in Figure 15.10. Perhaps this may prove impossible, but even at the seeing limit, this telescope will see objects ten times dimmer than is possible using other telescopes, so there will still be many unexpected discoveries to be made. After consulting several potential vendors, the OWL team believes that a realistic cost estimate is around €1–1.5 billion. With funding approval, first light for the full telescope would be sometime around 2021, with a partially completed instrument operating a few years before then. While this may seem a large cost (okay, it *is* a large cost), it should be noted that it is still less than the initial cost of the Hubble Space Telescope. Ultimately, however, the cost may still be too great, so OWL may have to be scaled down to a 'mere' 50-m version. Still, there can be no doubt that we have come a long way in the last 400 years.

Figure 15.10: A comparison between an 8-m seeing-limited ground-based telescope, the Hubble Space Telescope, an 8-m diffraction-limited telescope and a diffraction-limited 100-m telescope. Courtesy: ESO.

One last word

The last two decades have seen a dramatic increase in the number of telescopes with a diameter of 4 metres or larger. Advanced computer control has made it possible to accurately drive two axes of a large structure with a high degree of precision. This in turn led to the rise of the altitude-azimuth mount with the centre of mass of a telescope located directly on top of the support pier instead of off to one side as was the case with the equatorial mount. A more stable mount then made it possible to construct much larger and heavier telescopes. On the optics side, the development of segmented primaries and advanced figuring and testing procedures have allowed us to create a high-quality mirror with just about any shape while still managing to keep the mass down. Lastly, we have the ability to 'fix' the wavefront using adaptive and active optical controls to produce images almost as good as those from a space-based observatory.

The rise of improved technology is not the only driver. The spectacular images of cosmic objects we get from telescopes all around the world have increased the fascination of the public at large with the subject. They have also dramatically improved our understanding of the birth, development and future of the universe. For the most part, the cosmos is empty space with a temperature only a few degrees above absolute zero. Conversely, the majority of what we see through our telescopes is extremely hot. Stars, nebulae, supernovae and even the light from near black holes come from fantastically violent processes. Positioned in the delicate middle ground between these two extremes, however, some moderate climes exist. The odd rocky ball can form, and with it, complex molecules and even life. There can be no doubting that we are a rarity in more than one sense. However, our discoveries have also helped demystify our existence to show that Earthlings are mere bit-part actors in a larger production. The universe no longer revolves around us, nor are our planet, Sun or even our galaxy singularly special.

We now have conclusive proof that the all of the matter we can see forms only 4% of the total mass of the universe, with 23% being in the form of cold dark matter and 73% dark energy. This means that we cannot see 96% of the 'stuff' the universe is made of. However, we can endeavour to find out more about the nature of this elusive material by making detailed observations of the way the visible matter is distributed and moves about. In doing so, physicists are trying to develop a consistent theory which can explain the nature of the universe. Such theories describe high-energy effects well beyond what we could ever hope to create with Earth-based particle accelerators. Only in high-energy cosmic phenomena such as black holes, neutron stars, quasars and gamma-ray bursters are we able to observe such processes taking place. Thankfully for life on Earth, none of these celestial fireworks lies in our neighborhood. The flip-side then becomes the requirement to have large telescopes to make precise observations of the properties and behaviour of such distant phenomena.

A more selfish interest is also at work in creating better observatories. While we have a good understanding of the universe as a whole, we still have many questions concerning the rise of life within it. In the past decade we have uncovered conclusive proof of the existence of planets around other stars. While we have finally taken our first image of such an exo-solar planet, it is certainly a gas giant like Jupiter which is most unlikely to support life as we know it. Our next step is to take our first snapshot of an Earth-like planet and to make some measurements of the nature of any atmosphere it may have. We hope these measurements will be sufficient to enable us to look for biomarkers – the chemical signatures of life. Of course, the ultimate goal of this research is to answer the number-one question: 'Are we alone?'

This is just some of what to expect from the next generation of telescopes, but the real reason for building them is the discoveries that we cannot anticipate. There is a long track record of such unforeseen findings. The phases of Venus conclusively demonstrated that the Sun was the centre of the Solar System. The parallax of the star 61 Cygni gave us our first measurement of the true vastness of space. The discoveries of new planets, the rings of Saturn, ice on Mars, and even the first images of the far side of the Moon were all unexpected. All of these discoveries relied on the vision of telescopes. And they all completely altered the way we view our universe and our place in it. Though I am taking his comments somewhat out of context, I believe that Descartes probably summed up best the reason for building telescopes when he wrote:

> There can be no truth so remote that we cannot reach it, or so hidden that we cannot discover it.

Appendix A
Some mathematical basics

It would be wrong to assume that the reader of this book has a complete grounding in all aspects of mathematics, so this section is aimed at providing some background that should aid understanding of some of the terms. More to the point, it includes some 'rules of thumb' that may help in relating them to our everyday lives. One more note: in this book the units are the International System of Units (SI), with a couple of notable exceptions: Celsius is often used instead of Kelvin for temperature, and often electron-volts are used instead of Joules for energy.

Orders of magnitude

This book deals with the very large and the very small, and as such scientific notation is used to encompass these huge ranges in magnitude. I also often refer to standard prefixes which relate to these numbers. Below is a table which summarises terms and numbers used in the book.

Prefix	Also called	Notation	Value
femto		10^{-15}	0.000 000 000 000 001
pico	trillionth	10^{-12}	0.000 000 000 001
nano	billionth	10^{-9}	0.000 000 001
micro	millionth	10^{-6}	0.000 001
milli	thousandth	10^{-3}	0.001

		10^0	1.0
kilo	thousand	10^3	1,000
mega	million	10^6	1,000,000
giga	billion	10^9	1,000,000,000
terra	trillion	10^{12}	1,000,000,000,000
peta	quadrillion	10^{15}	1,000,000,000,000,000

Table A.1: A list of terms and descriptions of mathematical magnitudes.

So, for example, a nanometre is 10^{-9} m or 0.000000001 m or a billionth of a metre. To confuse things even further, the terms billion and trillion can have different meanings in different countries. In this book, they have the values assigned in Table A.1.

Angles

Telescopes are all about magnifying objects, which naturally means that what appears to subtend a small angle to the unaided eye becomes much larger when viewed through a telescope. In this book there is a great deal of discussion involving angles and there are two main conventions used: degrees and radians. A full circle can be divided up into 360 degrees (360°) or 2π radians (= 6.283185307...). In much of this book we will be concerned with very small angles, so let's see how these get divided up:

One full circle = 360 degrees
1 degree = 60 arcminutes
1 arcminute = 60 arcseconds
So there are 60×60×360=1,296,000 arcseconds in a full circle.

Radians are simpler; if we want one hundredth of a full circle, we simply divide 2π by 100 (= 0.06283...). But why do we need to use another type of scale at all? Well, it is useful as we can get the length of an arc by simple geometry as shown in Figure A.1, where distance (l) multiplied by the angle in radians (α) gives the size of the object (x):

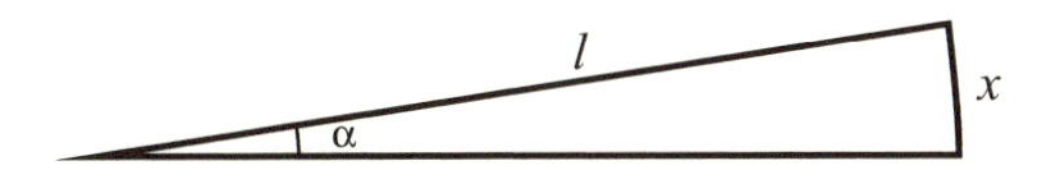

Figure A.1: Simple geometry shows that $x=l\alpha$, for α in radians.

So when making calculations, we prefer to use radians as the mathematics is a lot easier, though we often convert to degrees, minutes and seconds at the end. Don't question it, it is simply convention. To convert radians into degrees, you simply multiply by 180 and divide by π. Alternatively, for a rough estimate, you can use the fact that 2 arcseconds is around 10 microradians. It may also help to have some rules of thumb to relate these numbers to the real world, so these are given below.

Angle	Rough real-world guide
20 degrees	A hand-span at arm's length
4×25 degrees	Angular size of Ursa Major (The Big Dipper)
4.5×6 degrees	Angular size of Crux (The Southern Cross)
3 degrees	Angular size of Orion's Belt
1.5 degrees	A thumbwidth at arm's length
0.5 degrees	The angular size of the Sun and the Full Moon
1 arcminute	Resolution limit of the human eye
0.25 arcminute	A human hair at arm's length
10–60 arcseconds	Angular diameter of Venus
4–25 arcseconds	Angular diameter of Mars
40 arcseconds	Average angular diameter of Jupiter
18 arcseconds	Average angular diameter of Saturn
0.2 arcseconds	A human hair a block (100 m) away
20 milliarcseconds	Angular diameter of Betelgeuse
5.5 milliarcseconds	Angular diameter of Pluto (now)

Table A.2: A list of angular magnitudes and how to relate them to everyday situations.

Small-scale distances

We are all familiar with millimetres, metres and kilometres, but we rarely go beyond these numbers in our daily lives. Since we are going to address the very small and the very large in this book, some help may be required to get a feel for their values. Astronomers use a variety of distances in order to make it (relatively) easy to discuss cosmic scales while keeping the numbers to manageable values. Given that the Andromeda galaxy is 274,000,000,000,000,000,000,000 metres from Earth, you can probably see why it might be easier to say '29 million light-years' instead.

On the small scale, in reductions of thousands, we have metres, millimetres, micrometres (or microns) and then nanometres. In this book we will often refer to nanometres and microns as they are the scales of light waves and surface features of importance for optical fabrication. The table below gives some feel for the sizes of objects on the microscopic scale.

Distance	Rough real world guide
1 femtometre	Diameter of a neutron or proton
0.1 nanometres	Diameter of a silicon atom
20 nanometres	Width of a strand of DNA
120 nanometres	Size of the human immunodeficiency virus (HIV)
400 nanometres	Wavelength of violet light
500 nanometres	Wavelength of green light
600–700 nanometres	Wavelength of red light
2 micrometres	Size of *Escherichia coli* bacterium
7 micrometres	Diameter of human blood cell
60–100 micrometres	Diameter of an average human hair

Table A.3: A list of small measurements.

Large-scale distances

An astronomical unit (AU) is the distance from the Earth to the Sun: around 150 million kilometres, or the distance light travels in a little over eight minutes. Astronomical units are used for distances on the scale of the Solar System. For example, Jupiter orbits 5.2 AU from the Sun, and Neptune at 30 AU. This is much easier than using comparative values of 780 billion metres and 4.5 trillion metres respectively.

A light-year (ly) is the distance that light travels in one year, or almost 10^{16} metres. Such a number is virtually impossible for the human mind to grasp, but let's try anyway. If the Earth were reduced to the diameter of a grain of sand, then a light-year would be a distance of 1000 km. If you're still having a tough time with this, don't be disheartened, these distances are called astronomical for a reason.

A parsec (pc) is 3.26 light-years and the name is a contraction of 'parallax arcsecond'. Hold a single finger vertically in front of your face. Close first one eye then the other and you will see it move from side to side in relation to objects in the distance. In the same manner, as the Earth moves in its orbit around the Sun, a nearby star may seem to move in relation to more distant background stars. For an angular motion of 1 arcsecond for half a complete orbit (1 AU baseline), a star would be 1 parsec away. In 1838, the German astronomer/mathematician Friedrich Bessel measured the parallax of 61 Cygni to be 0.314 arcseconds, which would place it a little over 10 light-years away. Coincidentally, his name pops up again as a result of the fact that the mathematical formulation to describe the shape of the Airy spot (Chapter 4) involves Bessel functions.

For truly large distances we can talk about billions of light-years or parsecs, but cosmologists often prefer redshift. This is only natural since we calculate the distances to these objects by measuring their redshift, which is given the symbol 'z'. Nearly all distant objects are receding from us due to the expansion of the universe. The rate of this expansion is related to the distance of the object and can be measured as a shift in the emitted spectrum of light towards the red. We can calculate the redshift by using the formula:

$$z = \frac{\lambda_{measured} - \lambda_{emitted}}{\lambda_{emitted}}$$

Once we have a value for redshift, we can convert to distance by using the Hubble Constant (H_o = 71 km per sec per Megaparsec) using the formula $d=c\,z/H_o$. However, this is an approximation which only really works when z is much less than 1. A more correct formula needs to take into account the shape of the universe, which we now know to be flat. This leads to the more correct (and complex) formula:

$$d = \frac{c}{H_o}\left(1 - \frac{1}{\sqrt{1+z}}\right)$$

So with all of this in mind, we can create another table which will (we hope) put some of this in perspective (sic):

Distance	Rough real-world guide
100 km	Edge of the Earth's atmosphere
200–1200 km	Low/Near-Earth Orbit
36,000 km	Geosynchronous/Geostationary Earth Orbit
385,000 km	Earth/Moon distance
150,000,000 km or 1 AU	Sun/Earth distance
4.2–6.2 AU	Sun/Jupiter distance
30–50 AU	Sun/Pluto distance
4.2 ly or 1.3 pc	Distance to nearest star (Proxima Centauri)
425 ly or 130 pc	Distance to Betelgeuse
26,000 ly or 8,000 pc	Distance to centre of Milky Way
160,000 ly or 50,000 pc	Distance to Large Magellanic Cloud
65 million ly	Distance to Virgo supercluster
6.9 billion ly or $z = 3$	Distance to most distant quasar
9.6 billion ly or $z = 10$	Distance to most distant object seen (galaxy)

Table A.4: A list of large distances.

Appendix B
Electromagnetic radiation

What is light?

While this may seem a fairly basic question, it has been something of a thorn in the side of physicists for centuries. The reason has to do with quantum mechanics, which says that, depending on how you look at the problem, light can either be a wave or a particle. Throughout this book I too have often referred to 'electromagnetic waves' or 'photons' as the two aspects of its nature. There is often no reason for selecting one over another except that I have endeavoured to use the most appropriate picture in order to simplify the understanding of the concept I am addressing at the time. For example, in imaging situations, we tend to talk about photon counting, while in interferometry it is simplest to think of how waves add together to generate the fringes we look at. Still, in either picture there are several things to know. We'll start with the picture of light as a wave, which was the prevailing view until a century ago.

Light can be a transverse wave, which means it looks like a lot like a water wave as shown in Figure B.1. When we talk about the colour of light, we actually mean the 'wavelength', given the Greek symbol λ (lambda). Often we shorten this simply to 'wave' when clearly referring to distance. The wavelength is the distance from one point on the wave to the next corresponding point on the wave – say, the distance from the crest to the next crest. In the visible spectrum, light goes from violet (short wavelengths of around 400 nm) to red (longer wavelengths up to 700 nm). Historically we have developed the view that there are seven colours of the rainbow. This is a fallacy, in the same way that the ancients believed that everything in the universe was made up of the basic 'elements' earth, wind, fire and water. In fact, there are infinitely many colours, each one with a slightly different wavelength. Still, we often use the terms red, blue, green, etc., as they helps us narrow our discussions down to parts of the spectrum. Basically it's a lot easier to say 'the Moon is blue' rather than 'the Moon is emitting light at wavelengths primarily in the range four hundred to four hundred and fifty nanometres'. Who says physicists can't be romantic?

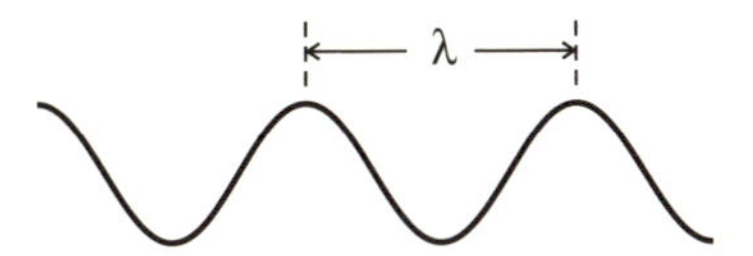

Figure B.1: This is a conceptual drawing of a light wave. The wave in this case could be moving to the left or the right, and has a wavelength (λ) as shown.

Being more specific now, we should really say that light is an electromagnetic wave. In 1873 a Scottish physicist named James Clerk Maxwell published a theory of electricity and magnetism which showed that they are directly linked to each other – when you have a varying magnetic field you create a varying electric field and vice versa. What was even more amazing, when you looked at an electric field travelling along (with its companion magnetic field), you can calculate the velocity of such a wave and it comes to the speed of light! So Maxwell inferred from this that in fact light is made up of sinusoidal oscillations of electric and magnetic fields at right angles to each other as shown in Figure B.2. Thus a more correct term for 'light' is 'electromagnetic radiation'.

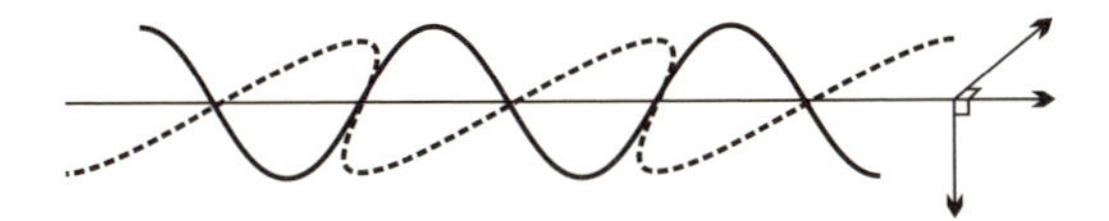

Figure B.2: An electromagnetic wave. Light consists of an electric field (dashed) and a magnetic field (solid) oscillating at right angles to each other.

Other forms of electromagnetic radiation appear on either side of the visible spectrum. For example, infrared (IR) radiation has a longer wavelength than our eyes are sensitive to, and still increasing wavelengths stretch towards microwave and the radio waves. On the other end of the visible spectrum we have ultraviolet (UV), which has wavelengths shorter than violet light. Then we have X-rays and

Gamma-rays (γ-rays). It is important to note that there is no real sense in dividing up the spectrum into these arbitrary regions; it is not as though there is any sudden change in the properties of these waves. In fact, the conventions mostly came about through historical discoveries made long before it was realised that these different emissions were all part of a related phenomenon. We have retained the naming conventions mostly because it is easier to restrict our discussions to certain parts of the spectrum.

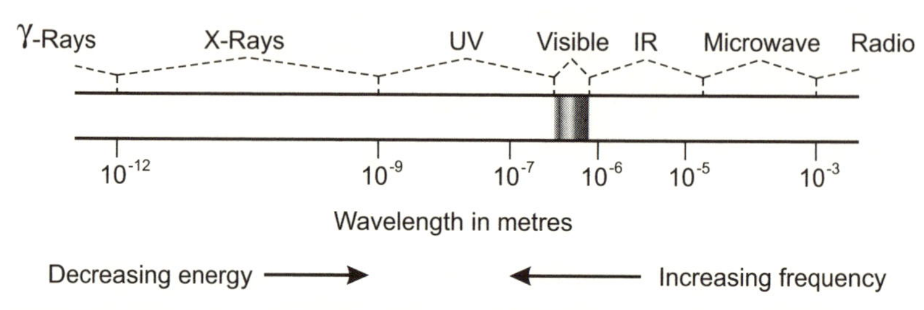

Figure B.3: This figure shows the different types of electromagnetic radiation along with the wavelengths. The visible regime occupies a very small portion of this spectrum.

One important thing to note about the entire electromagnetic spectrum is that the energy of the emission is related to the wavelength. At longer wavelengths, the energy of the wave is lower than at shorter wavelengths. Thus X-rays are higher-energy waves than ultraviolet rays which, in turn, have more energy than visible light. This makes sense when we think of how careful we have to be about exposure to X-rays and UV as they can have damaging effects on the body, whereas radio waves are around us all the time and they don't harm us.[40] Another way you can relate this to real life is by thinking about heating something. Things that are somewhat hot have a reddish glow (such as burning embers), but as we increase the heat we get a bluer colour of emission (such as a blow-torch). Physicists have equations (naturally!) to describe how a so-called blackbody emits electromagnetic radiation depending on the amount of heat it has. An object at around 5000 degrees Celsius will have peak emission at orange-yellow – such as our Sun. Meanwhile humans with an approximate temperature of 37 °C glow very brightly at infrared wavelengths of ~10 micrometres. For surveillance systems, this is significant, since this would be the optimum wavelength for spy satellites to use to better see people at night.

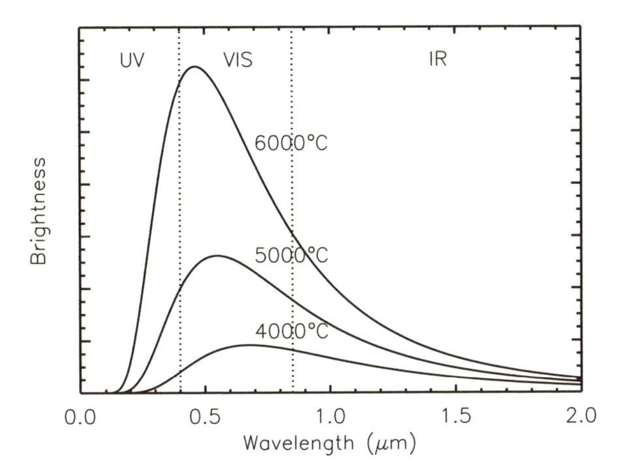

Figure B.4: This figure shows the curve for blackbody radiation. Notice the peak emissions wavelength for a lower-temperature object is towards the red end of the spectrum.

It was in trying to understand the blackbody radiation curve that Planck and Einstein theorised that light can come in discrete bundles known as photons. What's more, the shorter the wavelength, the more energy is contained in each photon. Of course, an individual visible photon has very little energy – some 4×10^{-19} Joules. In other words, it would take three billion billion of them to heat a thimble of water a single degree. An alternative view (so to speak) is to consider a moderately bright star in the night sky. Every second or so, a few million photons are being detected by your retinal rods and cones in order for you to see it. This leads us to stellar magnitudes, and the way in which we describe the brightness of stars.

Magnitudes

In order to have some sort of scale in which to set the relative brightness of particular objects in the night sky, scientists have adopted a (rather confusing) scale, known as apparent magnitudes. It is a system cobbled together from the time when Hipparchus, a Greek astronomer, decided to set down a standard for such things. Unfortunately, star brightness is a subjective issue when dealing with naked-eye observations, so the modern-day definition has altered things somewhat. In basic terms, the apparent magnitude (m) of Vega is set at 0. For every 2.512 times increase in brightness, the apparent magnitude decreases by 1. Thus the brightest star in the sky (Sirius) is magnitude –1.5 which means it is $2.512^{1.5} = 4$ times brighter than Vega. The list in Table B.1 should help set some benchmarks for how the scale works.

Object	Apparent magnitude
Sun	−26.8
Full moon	−12.5
Venus (max)	−4.4
Mars (max)	−2.8
Jupiter (max)	−2.7
Sirius	−1.5
Alpha Centauri	0
Vega	0
Betelgeuse	0.45
Saturn (max)	1
Polaris (variable)	2
Visual limit	6
Proxima Centauri	11
Brightest quasar	12.6
Pluto	15.1
Ground telescope limit	27–28
HST limit	30

Table B.1: A list of some common objects and their visual apparent magnitudes.

Generally the human eye can detect stars as faint as an apparent magnitude of 6 or so. Some astronomers have apparently demonstrated an ability to detect stars of magnitude 8.5, some ten times fainter. I have often doubted this, but then people quote the 'well-known' fact that the human eye is able to detect a single photon of light. This number is actually based on the fact that experiments have shown that a single rod can give a measurable response to a single photon. In fact, a rod in the human retina requires 5–10 photons in 0.1 seconds to set off the, 'Hey, there's some light there!' signal from the brain. Given that seeing a magnitude 8.5 star would require the detection of about 35 visible photons within a thirtieth of a second, this may not be unreasonable. At any rate, it is certain that we possess an incredible device for faint-object viewing.

Note: in the modern definition, astronomers have to make allowances for the colour of the star. After all, we need a way to differentiate between a blue and a red star. To take this into account, the magnitude is based on the brightness at a couple of 'bands' of wavelengths, but when quoted as a single number it is usually taken to be in the green. In the following section we will look at these bands in more detail.

Photometric bands

Astronomers divide the spectrum up into photometric bands, which aid in multicolour imaging and simply help in determining a standard portion of the spectrum over which observations are taken. For example, it is easy to understand the phrase 'we observed the star in the blue part of the spectrum', but for observations in the infrared, astronomers need a similar terminology. Often you will hear the phrase 'K-band observations of star umpty-ump showed ...'. For this, a series of bands have been set to divide the spectrum into more convenient, standardised chunks. The alphabetic designation of the band, the centre wavelength, and the size of each band (known as bandwidth) are given in Table B.2.

Filter band	Wavelength (μm)	Bandwidth (μm)
U	0.365	0.068
B	0.44	0.098
V	0.55	0.089
R	0.7	0.22
I	0.9	0.24
J	1.25	0.34
H	1.65	0.38
K	2.2	0.6
L	3.6	1.2
M	4.8	0.8
N	10.2	5.0

Table B.2: A list of properties for some visible and infrared photometric bands.

Energy

There are two units of energy used in this text. Joules are the correct international standard, but they only make sense in our everyday lives. As a result of the historical progression of discoveries and usage of various parts of the spectrum, we find that scientists often use different terms. Radio and microwave astronomers tend to talk about frequency, while infrared, visible and ultraviolet observations are often described in terms of wavelength. Meanwhile, at the high-energy end of the spectrum, we tend to use electron-Volts to describe X-rays and gamma-rays. One electron-Volt (eV) is equal to 1.6×10^{-19} Joules, and a 10 MeV gamma-ray would have a wavelength of 120 femtometres, or about 1000th the diameter of an atom.

Appendix C
Getting your own telescope

There's some hope that this book has sparked an interest in astronomy or telescopes in general and you are now considering purchasing or constructing a telescope. Before you do so, however, it is a good idea to know why you want this telescope, what your budget is, and what you hope to get out of it. If the telescope is to be used for observing from the city or looking through the neighbour's windows, then you won't need a particularly good one. If you can get out of populated areas to dark locations, it will be worth spending a little more to make the most of the abundant sights. If you can only spend $200, but expect to look through the eyepiece from a suburban backyard and see images much like those captured by the Hubble Space Telescope, you should be ready for some disappointment. Still, with little effort and expense, you can see some impressive sights. Here are some tips to make the selection process easier.

Aperture is everything

Until you get into the most expensive systems, the most significant factor is the diameter of the primary. A larger aperture means more collected light. Much like turning up the lights in a dark room, you can expect to see an amazing array of wonderful sights simply by making them brighter. Also a little gain in size goes a long way since brightness improves as the diameter *squared*. Below 5 cm (2 inches) it's not worth the effort, so an aperture of 10–20 cm (4–8 inches) is a good place to start. Telescopes with these diameters will be light and easy to take on trips to dark sites and they can be made well for little cost. Just remember that the optics must be high quality or all bets are off.

Refractor vs reflector

Unless you are willing to dip into your kid's college fund on your new hobby, a reflector is the way to go. A diffraction-limited, 15-cm (6-inch) reflector is quite affordable, but a high-quality refractor of the same size will cost many thousands of dollars. A cheap refractor is a waste of money as it will give blurry images with unsharp colours that will only disappoint. Reflectors are usually Schmidt-Cassegrain types (which are compact) and have wide fields of view and minimal chromatic aberration. They are also better suited (or at least more often designed) for photography. A Newtonian will be cheaper but more cumbersome and less likely to be set up for tracking objects or taking photos.

Magnification

Do not buy into the magnification hype. A large magnification is useless unless you have a very stable mount, since image shake will blur everything. Besides, once the aperture is larger than around 10 cm (4 inches), atmospheric blurring will put limits on the smallest details which can be seen. So no advantage will be gained from magnifying *ad infinitum*. Also, some of the best objects are fairly close, so you will want to make sure the field of view is wide enough to take them in and see them in all their glory. Magnify too much and you will make it hard to see the forest for the trees.

Mounting

Even the smallest telescope will have a significant magnification, and the slightest vibration will be magnified by the same amount. For this reason, you will need a sturdy mount to make sure objects don't move about. These days, tripod mounts come with fairly inexpensive drive systems and are simple to set up. The drive has the advantage that once an object is in view, you won't have to fiddle with knobs to make sure it stays that way. Most drive mounts will also have the capability to 'pick and point'. That is, you can choose an object from a list, press a button and the telescope will steer straight to that object (and begin tracking once it gets there). This is a great idea if you are not too familiar with all that the night sky has on offer.

Accessories

You should never look at the Sun without the protection of a filter over the end of the telescope. Irreparable damage to the eye will result. However, if you get a solar filter you can safely look at the Sun and see such things as sunspots and flares. You can also watch transits of Mercury and (more rarely) Venus. When these planets pass in between the Sun and the Earth, they appear as small black dots moving slowly across the solar disk. Choose the filter carefully, as some can darken the visible light but still let through harmful ultraviolet rays.

If you plan to take photos through the telescope, then getting one with a camera adaptor is a good idea (even if you don't use it). These can take either film cameras or CCD imaging systems. Of course, a camera adaptor only makes sense if you can track objects, so don't get one unless you have a tracking mount.

Price estimates

A rough estimate of price is about US$800–1000 for a 12-cm (5-inch) telescope and $1500–2000 for a 20-cm (18-inch) one. These will include all the mounts, electronics and accessories (except for a camera). That being said, they should last a long time and give you years of enjoyment.

A word of caution

From the time you first look through your telescope there is a risk of a lifelong addiction for which there is no cure. Robert Frost, an amateur astronomer himself, knew this well and wrote a poem called 'The Star-Splitter'. He tells the story of a farmer, Brad McLaughlin, who becomes a slave to his passion:

> *He burned his house down for the fire insurance*
> *And spent the proceeds on a telescope*
> *To satisfy a lifelong curiosity*
> *About our place among the infinities.*

Notes

1 For more on wavelengths, distances and angles, see Appendices A and B.

2 Tycho is the Latinised form of his true name, Tyge. The correct pronunciation of Tycho Brahe is *Too-co Bra*, but the name is often pronounced as something more like *Tie-co Bray* by most astronomers.

3 This historical record also seems to anticipate the telescope's use as a voyeuristic medium. Of course, the 'Peeping Tom' mentioned here came some five hundred years after the phrase was coined in the story of Lady Godiva.

4 It actually depends on the square root of the length and the inverse square root of the local acceleration due to gravity, which is pretty much constant over the surface of the Earth.

5 Furthermore, over the next few years Galileo went on to develop techniques to improve the quality of the lenses themselves for even better imaging.

6 It wasn't until 1847 that a naming convention for the satellites of Jupiter which had been suggested by Kepler two centuries earlier was generally agreed upon: the names of the loves of Jupiter from mythology. Now collectively referred to as the Galilean moons, the names of the satellites (from inner to outer) are Io, Europa, Ganymede and Callisto.

7 This leads to many philosophical questions, the most profound of which is: can the image seen through a telescope be considered real?

8 Pronouncing this surname correctly is all but impossible for those who do not speak Dutch. An adequate compromise adopted by most physicists is *Hoy-genz*.

9 From the same word in Latin meaning 'mirror'.

10 In actual fact, both the Cassegrain and Gregorian designs were

conceived in 1636 by a French mathematician and monk, Marin Mersenne. His designs were not well developed and, as far as we know, he never tried to make one.

11 Mars does have large canyons and channels, some of which are in fact larger than the Grand Canyon, but these would not have been visible to either astronomer.

12 The image resolution was actually larger than the factor of two that these numbers would suggest, owing to improvements in digital recording methods. These issues will be discussed in later chapters.

13 Actually, today the quarter wavelength criterion is rarely used. Instead, astronomers have more informative methods of characterising a wavefront (such as root mean square error and Strehl ratio). These tend to be fairly complex, so I will stay with this 'old-fashioned' definition as it is still a good rule of thumb.

14 In fact, in response to the increase in light levels, your iris will close, which *reduces* image resolution. Meanwhile, by using the cones in the fovea instead of the widely space rods, you will get an *increase* in resolution. Overall there is little appreciable change.

15 Astronomers still tend to go to the observatory during their observations, with the typical reason given that it helps to be on site in case of problems. Given the large support staff and access to real-time communications that makes this unnecessary, I tend to believe the more obvious reason – it's a great excuse to go to Hawaii and Chile.

16 Universal Time (UT) is the more correct usage of what was formerly known as Greenwich Mean Time (GMT).

17 This is not the case for all portions of the sky, since the Moon, Earth and Sun need to be avoided.

18 Interestingly, the closest known Cepheid to the Earth is actually Polaris. Thus, the well known aphorism, 'As constant as the North Star' couldn't be further from the truth. It is constant neither in location (because of precession) nor in brightness.

19 Perkin-Elmer was the original name of the company at the time of fabricating the Hubble Telescope mirror. The optical fabrication division has since changed hands several times and is now owned by Goodrich. From this point onwards I will refer to the company by its former name.

20 More precisely, the conic constant was made to be –1.0139 instead of –1.0023.

21 We now know from subsequent measurements (including another NASA probe: the Wilkinson Microwave Anisotropy Probe), that H_o = 71 km/s/Mpc, giving an age to the universe of 13.7 billion years.
22 My personal belief is that it was renamed because the contract for its construction went to Northrop Grumman, which may have led many people to suspect an alternative explanation behind the NGST acronym.
23 Strictly speaking, gravitational bending should only change with angle above the horizon (altitude/elevation angle) and not compass direction (azimuth angle), which simplifies things greatly.
24 Note that this number of subapertures was calculated for operation in the infrared where r_o is much larger.
25 To be precise, the mirror is given the shape of half the deformation, since the net wavefront correction will be doubled in the process of reflection.
26 As it turns out, astronomers using the Hubble Space Telescope had taken images of the separated binary stars nine months earlier, but did not publish these results until mid-1997.
27 In actual fact, most 'optical' fibres actually use infrared wavelengths of 850 nm, 1310 nm or 1550 nm, so the term is something of a misnomer.
28 Twice, since it went out and then came back.
29 Strictly speaking, Rayleigh lidar measures relative temperature and needs to be calibrated to a known temperature, but the end result is the same.
30 At the risk of stating the obvious, it is perhaps worth noting that places such as these probably have sensors to detect laser illumination. In other words, it is best not to try this unless you want an unpleasant visit from the Secret Service.
31 In fact the benefits were anticipated as far back as 1946, in a RAND study entitled: *Preliminary Design of an Experimental World-Circling Spaceship.*
32 The very existence of the NRO itself was not publicly acknowledged until 1992.
33 While you may expect ultraviolet to improve the resolution still further (and you'd be right), the atmosphere is opaque at such wavelengths. Going further down the spectrum wouldn't help anyway since most objects barely radiate at such wavelengths.
34 There is no violation of relativity here, as the rays are travelling

faster than the speed of light in the air but not faster than the speed of light in a vacuum.

35 The term 'photon' was not actually coined until 1928, but we can use this current term for the quantized nature of light that Einstein and his contemporaries were introducing.

36 Hence the use of the term 'Special', as the theory pertains to the specialised case where there is no acceleration.

37 Note that we are not breaking any laws of physics here by destroying energy. Just as in the case of the nulling interferometer in Chapter 7, the LIGO interferometer has two output ports, so all the light goes out of the other one. This light is then recycled back into the interferometer to improve its sensitivity.

38 This is around the distance to the Virgo supercluster – a large conglomeration of some 5000 galaxies.

39 In order to get an idea of this extreme aspect ratio, place two compact disks on top of one another. A marble placed in the centre would be the size of the nucleus bulge.

40 Up to a point at least. We can be harmed by all types of radiation if there is enough of it around. This is why we should avoid strong sources of microwave and infrared radiation as, in high enough fluxes, they can cook human flesh.

Bibliography

Books

F. Adams and G. Laughlin, *The Five Ages of the Universe*, The Free Press, 1999.
P. Y. Bely (ed.), *The Design and Construction of Large Optical Telescopes*, Springer-Verlag, 2003.
B. Berman, *Secrets of the Night Sky*, HarperCollins, 1995.
W. E. Burrows, *Deep Black*, Random House, 1986.
B. W. Carroll, *An Introduction to Modern Astrophysics*, Addison-Wesley, 1996.
E. J. Chaisson, *The Hubble Wars*, HarperCollins, 1994.
J. Cornell and A. P. Lightman (eds.), *Revealing the Universe*, MIT Press, 1982.
J. Cornell and J. Carr (eds.), *Infinite Vistas*, Charles Scribner's Sons, 1985.
R. P. Crease, *The Prism and the Pendulum*, Random House, 2003.
K. Croswell, *The Universe at Midnight*, The Free Press, 2001.
D. DeVorkin, *Beyond Earth*, National Geographic, 2002.
K. Ferguson, *Measuring the Universe*, Walker and Company, 1999.
T. Ferris, *Seeing in the Dark*, Simon and Schuster, 2002.
G. Galilei, *Sidereus Nuncius*, University of Chicago Press, 1989.
O. Gingerich, *The Book That Nobody Read*, Walker and Company, 2004.
J. Gribbin, *The Birth of Time*, Weidenfield and Nicholson, 1999.
J. Gribbin, *Science: A History*, Penguin Books, 2002.
A. W. Hirshfeld, *Parallax*, Henry Holt and Company, 2001.
T. Hockey, *Galileo's Planet*, Institute of Physics Publishing, 1999.
J. B. Kaler, *Stars*, Scientific American Library, 1992.
M. Katzman (ed.), *Laser Satellite Communications*, Prentice-Hall, 1987.
R. Kerrod, *Get a Grip on Astronomy*, Time-Life Books, 1999.
H. C. King, *The History of the Telescope*, Dover Publications, 1955.
D. Leverington, *New Cosmic Horizons*, Cambridge University Press, 2000.
J. E. Lewis, *Spy Capitalism*, Yale University Press, 2002.
M. Littmann, *Planets Beyond*, Dover Publications, 2004.
M. S. Longair, *High Energy Astrophysics*, Cambridge University Press, 1981.
L. J. Ludovici, *Seeing Near and Seeing Far*, John Baker Publishers, 1966.
D. Malacara (ed.), *Optical Shop Testing*, 2nd ed., John Wiley & Sons, 1992.
E. Maor, *June 8, 2004 – Venus in Transit*, Princeton University Press, 2000.
J. Meeus, *Astronomical Algorithms*, Willmann-Bell, 1991.
A. A. Michelson, *Studies in Optics*, Dover Publications, 1995.
I. Newton, *Opticks*, Dover Publications, 1952.
R. Panek, *Seeing and Believing*, Penguin Books, 1999.
A. Parker, *In the Blink of an Eye*, Basic Books, 2003.
B. Parker, *Stairway to the Stars*, Perseus, 2004.
M. Pendergrast, *Mirror Mirror*, Basic Books, 2003.
R. Preston, *First Light*, Abacus Books, 1987.
I. Ridpath (ed.), *Norton's 2000.0*, 18th ed., Longman Group, 1989.
F. Roddier (ed.), *Adaptive Optics in Astronomy*, Cambridge University Press, 1999.
V. Ronchi, *Optics*, Dover Publications, 1991.
C. Sagan, *The Demon-Haunted World*, Ballantine Books, 1996.
F. Schaaf, *The Starry Room*, John Wiley and Sons, 1988.
D. J. Schroeder, *Astronomical Optics*, Academic Press, 1987.
H. Shapley, *Beyond the Observatory*, Charles Scribner's Sons, 1967.

J. A. Simpson and E. S. C. Weiner (eds.), *Oxford English Dictionary*, 2nd ed., Clarendon Press, 1989.
D. Sobel, *Longitude*, Fourth Estate, 1995.
D. Sobel, *Galileo's Daughter*, Penguin Books, 2000.
R. Solé and S. Valbelle, *The Rosetta Stone*, Four Walls Eight Windows, 2002.
T. Standage, *The Neptune File*, Penguin Books, 2000.
P. Strain and F. Engle, *Looking at the Earth*, Turner, 1992.
H. R. Suiter, *Star Testing Astronomical Telescopes*, Willmann-Bell, 1994.
R. K. Tyson, *Principles of Adaptive Optics*, 2nd ed., Academic Press, 1998.
R. K. Tyson and B. W. Frazier, *Field Guide to Adaptive Optics*, SPIE Press, 2004.
A. Upgren, *Night has a Thousand Eyes*, Perseus, 1998.
J. R. Voelkel, *Johannes Kepler and the New Astronomy*, Oxford University Press, 1999.
F. Watson, *Stargazer*, Da Capo Press, 2004.
A. Waugh, *Time*, Headline, 1999.
R. N. Wilson, *Reflecting Optics*, Vols. I & II, Springer-Verlag, 2001.
G. Yost, *Spy-Tech*, Facts on File Publications, 1985.
M. V. Zombeck, *Handbook of Space Astronomy and Astrophysics*, 2nd ed., Cambridge University Press, 1990.

Journals, conference proceedings and other sources

Applied Optics
Astronomy
Astrophysical Journal
Astrophysical Journal Letters
BBC News
British Library
Geology
Istituto e Museo di Storia della Scienza
NASA Technical Reports
Nature
Science
Sky and Telescope
SPIE Proceedings, Vols. 3126, 3352, 3353, 3356, 3757, 3762, 4013, 4035, 4494, 4840, 4849, 5487, 5489, 5553
SPIE Milestone Series, Vols. MS73, MS93, MS139

Websites

ATST: atst.nso.edu
Airborne Laser: www.boeing.com/defense-space/military/abl
Anglo-Australian Observatory: www.aao.gov.au
Astronomy Picture of the Day: antwrp.gsfc.nasa.gov/apod/astropix.html
British Library: www.bl.uk
British Museum: www.thebritishmuseum.ac.uk
Californian/Carnegie Planet Search: exoplanets.org
Carnegie Observatories: www.ociw.edu
Cassini/Huygens Probe: saturn.jpl.nasa.gov/home/index.cfm
Center for Adaptive Optics: cfao.ucolick.org
David Malin Images: www.davidmalin.com
Digital Globe: www.digitalglobe.com
Digitized Sky Survey: archive.stsci.edu/dss
Dutch Open Telescope: dot.astro.uu.nl

European Southern Observatory: www.eso.org
European Space Agency: www.esa.int/esaCP/index.html
Federation of American Scientists: www.fas.org/main/home.jsp
Gemini Telescope: www.gemini.edu
Giant Magellan Telescope: www.gmto.org
Google Earth: earth.google.com
Gran Telescopio Canarias: www.gtc.iac.es/home_s.html
Heavens Above: www.heavens-above.com
HESS Homepage: www.mpi-hd.mpg.de/hfm/HESS/
Hobby-Eberly Telescope: www.as.utexas.edu/mcdonald/het/het.html
Hubble Heritage Project: heritage.stsci.edu
IAU Minor Planet Center: cfa-www.harvard.edu/cfa/ps/mpc.html
ImageSat: www.imagesatintl.com
Istituto e Museo di Storia della Scienza: galileo.imss.firenze.it
James Webb Space Telescope: www.jwst.nasa.gov
Keck Telescope: www2.keck.hawaii.edu
Kepler Probe: www.kepler.arc.nasa.gov
Large Binocular Telescope: medusa.as.arizona.edu/lbto/
Large Synoptic Survey Telescope: www.lsst.org/lsst_home.shtml
LIGO: medusa.as.arizona.edu/lbto/
Malin Space Science Systems: www.msss.com
Mount Wilson Telescope: www.mtwilson.edu
NASA Homepage: www.nasa.gov/home/index.html
National Reconnaissance Office: www.nro.gov
NOAA: www.noaa.gov
ORBIMAGE: www.orbimage.com
OWL Telescope Project: www.eso.org/projects/owl/
Palomar Observatory: www.astro.caltech.edu/palomar
Pan-STARRS: pan-starrs.ifa.hawaii.edu
Rice University Galileo Project: galileo.rice.edu
Royal Society: www.royalsoc.ac.uk
Royal Swedish Academy: www.solarphysics.kva.se
Sky and Telescope Webpage: skyandtelescope.com
SkyView Virtual Telescope: skyview.gsfc.nasa.gov
Sloan Digital Sky Survey: www.sdss.org
Smithsonian/NASA Abstracts: adsabs.harvard.edu/abstract_service.html
South African Large Telescope: www.salt.ac.za
SOHO: sohowww.nascom.nasa.gov
Space Imaging: www.spaceimaging.com
Spot Image: www.spotimage.fr
STSci Hubble Website: hubblesite.org
SEDS: seds.lpl.arizona.edu
TerraServer: terraserver.homeadvisor.msn.com
Terrestrial Planet Finder: planetquest.jpl.nasa.gov/TPF
Thirty Meter Telescope: www.tmt.org
UBC Liquid Mirror Page: www.astro.ubc.ca/LMT/index.html
UK Schmidt Telescope: www.aao.gov.au/ukst
United States Naval Observatory: aa.usno.navy.mil
Wikipedia: www.wikipedia.org

Index